Design Considerations for Space Elevator Tether Climbers

International Space Elevator Consortium

Fall 2013

Peter A. Swan
Cathy W. Swan
Robert E. 'Skip' Penny, Jr.
John M. Knapman
Peter N. Glaskowsky

A Study for Progress in Space Elevator Development

Design Considerations for Space Elevator Tether Climbers

Copyright © 2014 by:

Peter A. Swan
Cathy W. Swan
Robert E. 'Skip' Penny, Jr.
John M. Knapman
Peter N. Glaskowsky

All rights reserved, including the rights to reproduce this manuscript or portions thereof in any form.

Published by Lulu.com

dr-swan@cox.net

978-1-312-09982-1

Cover Illustration: chasedesignstudios.com

Printed in the United States of America

Preface

The International Space Elevator Consortium is a non-profit organization consisting of volunteers with the following vision and mission:

Vision: ***A world with inexpensive, safe, routine, and efficient access to space for the benefit of all mankind.***

Mission: ***The International Space Elevator Consortium promotes the development, construction and operation of a Space Elevator infrastructure as a revolutionary and efficient way off Earth for all humanity.***

ISEC pushes the knowledge of space elevators with multiple approaches. Each year the Board of Directors selects a topic to focus the efforts of the many volunteers. This leverages the various strengths of individuals while combining their strengths in team efforts. This selection of a topic focuses activities within the following;

- ISEC's annual journal, *Climb*
- The International Space Elevator Conference
- Comprehensive studies
- Study reports
- Monthly newsletter

Recent and planned topics include:

  2010 – Space Elevator Survivability – Space Debris Mitigation
  2011 – Carbon Nanotube developments
  2012 – Space Elevator Concept of Operations
  2013 – Tether Climbers
  2014 – Space Elevator Architectures and Roadmaps

The subject selected for this 2013 study is the Space Elevator Tether Climber. The objective of the one year study, conducted by ISEC with help from external sources, was to survey current concepts and technologies related to tether climbers, identify critical issues, questions, and concerns, assess their impact on the development of space elevators, and project towards the future. Two aspects of this work were especially important:

- Involvement of the space elevator community (especially at the International Space Elevator Conference), and
- Completion within one year to allow the team to move onto the next yearly theme and to enable sharing of the information.

The authors of this report wish to thank:

- The members of ISEC for their support
- Authors of the report for the dedicated efforts
- The attendees of the 2013 International Space Elevator Conference, and
- These peer reviewers for their insight into the technological topics: Harold S. Rhoads, Ph.D., Daniel Smolski, and Paul Phister, Ph.D.

Signed: *Peter A. Swan*,
President ISEC

Executive Summary

The 2013 ISEC study report addresses a critical aspect of the space elevator infrastructure: the tether climber. The tether climber will leverage 60 years of spacecraft design while incorporating aspects of traditional terrestrial transportation infrastructure. The following conclusions were refined from discussions during the year and the assessment of relevant research:

- The study used a constant-power model as a baseline, rather than constant speed or acceleration, because the constant-power model simplifies design requirements and reduces the mass of the tether climber.
- A mass of 6 metric tons (MT) for a climber and 14 MT for customer payloads seems feasible. We estimate that with a travel time of one week to Geosynchronous Earth Orbit (GEO), seven tether climbers can be on a 31 MT tether simultaneously.
- A conceptual operating plan should be developed to assist in the refinement of requirements.
- The communication architecture should be integrated into the space elevator infrastructure and nodal layout. This will enable the tether climber to be in constant contact with operators and customers.
- It appears possible to operate the tether system exclusively on solar power, eliminating the need for ground-level power sources.
- Exclusive use of ground-based laser power transmission also seems practical.
- A hybrid of solar power and laser power transmission is an option.
- Detailed designs of the climber, tether attachment apparatus, and drive mechanisms will be accomplished later when the characteristics of the tether are better defined, but it is clear that terrestrial motor designs can be leveraged for climber development.

The study report makes the following recommendations related to space elevator development:

1. Conduct further studies to lower technological risk in several specific areas (see section 5.3).
2. Continue to support technical competitions in areas such as tether design and tether climber design.
3. Develop more international connections during ISEC studies, especially with Japanese Space Elevator Association, EuroSpaceward, and the International Academy of Astronautics.

To accomplish this year-long study, the authors leveraged inputs from multiple sources, and would like to thanks the following innovative thinkers: Brad Edwards, Eric Westling, Ben Shelef, David Raitt, Cathy Swan, John Knapman, Skip Penny, Ted Semon, Larry Bartoszek, and Martin Lades.

Table of Contents

Preface iii
Executive Summary v

Chapter 1   Introduction 1
   1.1   Study process 1
   1.2   Previous studies 2
   1.3   Spacecraft design criteria 2
   1.4   The constant power baseline 4
   1.5   Customer requirements 4
      1.5.1   Nuclear waste disposal 5
      1.5.2   Earth sunshade 5
      1.5.3   Space-based solar power 5
   1.6   Spaceport locations—Where to park? 6
   1.7   Phases of ascent and descent 7
   1.8   Tether Climber Description 8
   1.9   Tether climber components 10
   1.10 Tether Climber Types 11

Chapter 2   Tether Climber Operational Phases 13
   2.1   Climber types and descriptions 14
   2.2   Phases of Operations 14
      2.2.1   Preparation for climb 15
      2.2.2   Attach 15
      2.2.3   Atmospheric lift (under 40 km) 16
      2.2.4   Heavy gear, high gravity 16
      2.2.5   High speed 17
      2.2.6   Detach at GEO node 17
      2.2.7   Coast to beyond GEO 17
      2.2.8   Load cargo 17
      2.2.9   Transfer to descending tether 17
      2.2.10   Attach to descending tether 17
      2.2.11   Descent to 40 kilometers 17
      2.2.12   Atmospheric re-entry 17
      2.2.13   Detach 18
      2.2.14   Transfer to ascending tether 18
   2.3   Climber Missions 18
      2.3.1   GEO Delivery and Pickup 18
      2.3.2   LEO Delivery 18
      2.3.3   Beyond GEO Delivery 18
      2.3.4   Where to Park? 19
      2.3.5   Return to Earth's surface 19
      2.3.6   Tether Repair 19
   2.4   Climber Operations Center 19
   2.5   Life Cycle of a Climber 20
   2.6   Operations growth potential 23

2.7	References	23

Chapter 3 Sub-System Description 24
- **3.1** Climber Types 24
- **3.2** Atmospheric Protective Shroud 25
- **3.3** Operational Tether Climber Design Basics 26
  - 3.3.1 Customer Needs 27
  - 3.3.2 Simultaneous Tether Climbers 28
  - 3.3.3 Constant Power Baseline 28
- **3.4** Tether Climber Sub-Systems 31
  - 3.4.1 Mass and Cost Savings Possible With Carbon Nanotubes 31
  - 3.4.2 Structure 32
  - 3.4.3 Energy Management 34
  - 3.4.4 Communications Payloads and Antennas 35
  - 3.4.5 Command and Data Handling (CADH) 35
  - 3.4.6 Environmental Control 35
  - 3.4.7 Attitude Determination and Control 36
- **3.5** Motor / engine and drive apparatus 37
  - 3.5.1 Wheel Interface Characteristics 38
  - 3.5.2 Friction 39
- **3.6** Tether Construction Climber – Initial Design 40
  - 3.6.1 The Mass Budget of the First Construction Climber and the Mass Problem of the 2004 Conceptual Design ... 43
  - 3.6.2 Conclusion on Construction Climber Design 46
- **3.7** Further Thoughts about Tether Climbers 46
  - 3.7.1 Panel A 47
  - 3.7.2 Panel B 47
  - 3.7.3 Panel C 47
  - 3.7.4 Panel D 48
- **3.8** Tether Climber Conclusions 48

Chapter 4 Power Sources 50
- **4.1** Laser Power Option 51
  - 4.1.1 Power receiver alignment 52
  - 4.1.2 Power receiver efficiency 53
  - 4.1.3 Engineering Reality Check 53
  - 4.1.4 Status 54
  - 4.1.5 Major Design Challenges 55
- **4.2** Solar Power Option 56
  - 4.2.1 Solar power array suspension 61
  - 4.2.2 Engineering Reality Check 62
  - 4.2.3 Laser Clearinghouse Requirements 63
- **4.3** Comparison of Power Options 64
  - 4.3.1 Laser Pros and Cons 64
  - 4.3.2 Solar Power Pros and Cons 64
- **4.4** The Hybrid Climber 65
- **4.5** Power Source Conclusions 67

Chapter 5 Conclusions and Recommendations 69

5.1	General Questions	69
5.2	Conclusions	70
5.3	Recommendations	70

Appendix A: Movement to 40 Kms 72

Appendix B: Tether Baseline 77

Appendix C: An Alternate Climbing Approach 79

Appendix D: The International Space Elevator Consortium 81

Chapter 1   Introduction

The International Space Elevator Consortium chooses a focus theme and conducts a study each year to address questions and issues relevant to the advancement of space elevators. This focus enables the organization to stimulate activities and products across a single topic and produce results that show up at the yearly conference, inside the organization's journal *CLIMB*, and during a year-long study producing a formal study report. The previous studies were:

    2010 – Space Elevator Survivability – Space Debris Mitigation
    2011 – Carbon Nanotube developments
    2012 – Space Elevator Concept of Operations

For 2013, the ISEC board of directors determined that a comprehensive look at the tether climber, a key element of the space elevator infrastructure, was needed. The tether climber shares some attributes of a traditional spacecraft while demanding special characteristics in operations and design. Some of the questions addressed by this study are:

- Do we know enough to design a basic tether climber?
- What special design characteristics are required?
- How will the tether climber interact with the tether?
- Can we meet the needs of the developer/owner/operator? (E.g., 6 MT climbers, 14 MT payloads, one launch per day, one-week trip to GEO, 7.5 year tether service life, 10 year MTBF)
- Which power source—solar, laser, or hybrid?
- Is the projected design achievable within the next 15 years?

The purpose of this report is to provide the space elevator community with a common starting point for further designs and projections of capability for tether climbers.

1.1   Study process
The study process was straightforward with the following steps:

August 2012	ISEC selects topic at Board of Directors meeting
	Tether climber announced as the topic at the yearly conference
Aug-Dec	Team formed and initial outline of study topics discussed
Jan-Mar 2013	Specific items discussed, analyzed and studied
Mar-Aug	Paper topics submitted to the ISEC International Conference
August	Focus at space elevator conference on tether climbers
	Mini-workshop brainstorming initiated tether climber ideas
Sep-Jan 2014	Study topics drafted as chapters in the report
Jan-Feb	ISEC Review of Final Document
Feb-Mar	Final review and top level peer review
April	Publish Study Report as ISEC STUDY

1.2 Previous studies

Past studies on space elevators have offered many concepts for tether climber design. These climber designs varied with respect to mass, shape, and power source. This ISEC Study has leveraged the excellent work of many engineers and scientists in their designs on tether climbers.

This study has permission to present many of their ideas and concepts. As the 2013 theme for ISEC, and the International Space Elevator Conference, many Tether Climber concepts were presented and have been included in this report. Dr. Knapman's paper on Constant Power [Knapman 2013] has become a baseline engineering approach for this report while Mr. Robert Penny [Penny, 2013] discussed the Operational Phases in his mini-workshop. Mr. Ted Semon [Semon 2013] presented the Hybrid Tether Climber, which will be presented in chapter 4.

The key papers, books, and presentations on tether climbers will be referenced during the study report, but they are also summarized in the list below:

- 2002, Dr. Edwards & Eric Westling, Space Elevators
- 2006, Ben Shelef, A Solar-Based Space Elevator Architecture (pdf report)
- 2013, International Academy of Astronautics Study Report, "Space Elevators: An Assessment of the Technological Feasibility and Way Forward."
- 2013, Robert "Skip" Penny, "Tether Climber Operational Phases," 2013 ISEC Conference, Seattle.
- 2013, Mr. Ted Semon, "Hybrid Tether Climber," 2013 ISEC Conference, Seattle.
- 2013, Dr. Knapman, "Tether Climber at Constant Power," 2013 ISEC Conference, Seattle.
- 2013, Mini-workshop Results, Tether Climber Brainstorming, 2013 ISEC Conference, Seattle.
- 2013, Bartoszek, Larry, "Getting the Mass of the First Construction Climber Under 900 Kg," 2013 ISEC Conference, Seattle.
- 2013, Lades, Martin, "Climber-Tether Interface for a Space Elevator, 2013 ISEC Conference, Seattle.
- http://spaceelevatorwiki.org
- http://en.wikipedia.org/wiki/Space_Elevator
- http://en.wikipedia.org/wiki/Wireless_power

1.3 Spacecraft design criteria

The tether climber is a spacecraft, and like other spacecraft, the tether climber will be an engineering marvel. Tether climbers will have many unique characteristics, starting with the method of propulsion—electric traction motors instead of rockets. From a passenger's perspective, the most unusual aspect of this spacecraft design is that a ride up or down a space elevator will be smooth and routine. Each day a new climber will begin its one-week trip to GEO without the "shake, rattle & roll" of rocket motors.

Design criteria will be based upon environmental requirements (vacuum, thermal, radiation, etc.) and user needs (size, shape, power, speed of climb, etc.). The design of the tether climber will leverage historical space designs, but relatively low aerodynamic loads enable superior design flexibility such as open sides to allow for unusually large payloads. The old restrictions of aero-shells and fairings will go away. Reduced launch loads eliminate the need for exceptionally strong structures in the climber and payloads.

Figure 1-1 shows the nodes of a space elevator infrastructure from the terminal station at the surface of the ocean (the *Marine Node*) to the counterweight approximately 100,000 km above (the *Apex Anchor*).

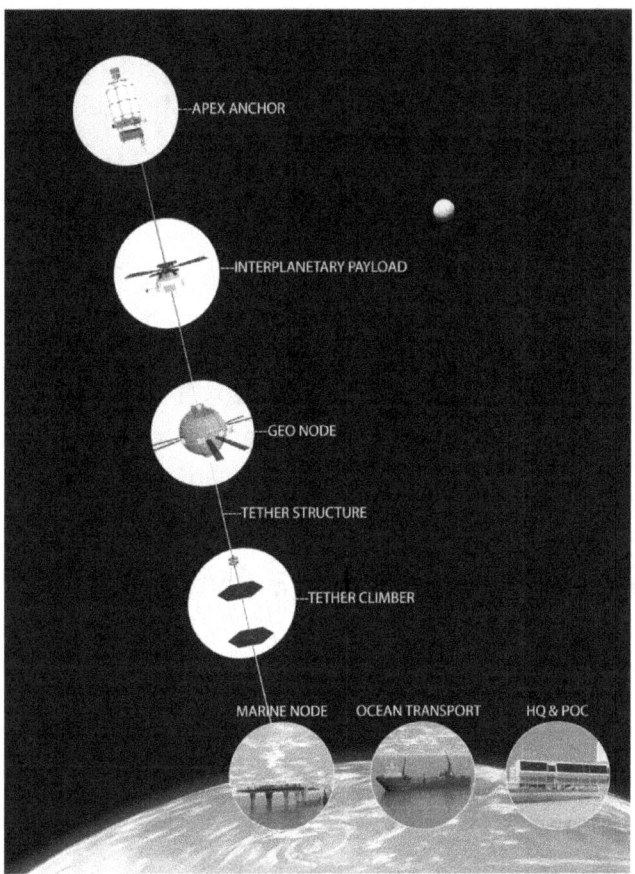

Figure 1-1. Space Elevator Nodes [Frank Chase Image]

Some of the technologies discussed in Chapter 3 (Tether Climber Systems Approach) refer to the increase in efficiencies of lightweight deployable solar cells, power projection capabilities of lasers, extremely strong and lightweight structures based upon carbon nanotubes (CNT), and phenomenal energy storage and distribution because of the new characteristics of CNT and battery technologies.

1.4 The constant power baseline

This study will address the full range of options for the space elevator tether climber. However, to enable the study to establish a design early in the concept development phase, the study proposes, as a baseline, a concept with tremendous strengths— the concept of *constant power*. This concept has been discussed previously and was presented at the 2013 International Space Elevator Conference. [Knapman 2013] The details of this concept, and its implications, will be expanded upon in Chapter 3 (Tether Climber Systems Approach).

In summary, constant power means that climbing speeds are lower in the early phase of the ascent, and increase at higher altitudes as the force of gravity declines. To achieve constant acceleration, the motor and tether gripper subsystems would have to be heavier—extra mass that would have to be carried throughout the lift. The constant-power solution also ensures that the full capacity of the laser and/or solar power sources (estimated at 4 megawatts) is utilized at all times.

1.5 Customer requirements

Systems engineering approaches for all large developments begin with the predicted customer requirements. Such predictions are somewhat arbitrary given that operations are twenty years away and the technologies are in the lower ranges of readiness levels. However, even a basic space elevator system would provide far more lift capacity to GEO than is currently available. The last ten years of ISEC's space elevator research provides a useful preliminary estimate of system capability:

- Revolutionarily inexpensive to GEO ($500/kg to GEO)
- Easy delivery to GEO within a week
- Commercial infrastructure development similar to bridge building
- Financial numbers that are infrastructure enabling
- Routine – daily launches
- Safe – no explosives, no free flight, low aerodynamic velocity
- Permanent – no disposable components
- Multiple space elevators when infrastructure matures
- Minimum 24/7/365 with 50-year lifespan
- Massive lift capacity – 14 metric tons of cargo per launch
- None of the shake, rattle, and roll of rocket launches
- More flexibility in cargo size and configuration
- Provides a convenient platform for recovery and repair of satellites
- Low environmental impact – solar power, no chemical propellants
- Leaves no space debris in orbit

According to an International Academy of Astronautics study [Swan 2013], a full space elevator system should include at least three pairs of space elevators distributed around the globe, operated as competing businesses, each providing daily lifts of up to 14 metric tons (MT) of customer payload.

This leads to an obvious question: what commercial payloads will need so much lift capacity? This ISEC report focuses on three missions for the future space elevator infrastructure as examples of future "breakout" businesses enabled by high-volume, low-cost access to space. (These descriptions are adapted from the IAA Cosmic Study [Swan, 2013].)

1.5.1 Nuclear waste disposal

Current plans for disposal of massive amounts of residual radioactive material across the world are being argued from many perspectives. The dominant argument is summarized by the catchphrase "not in my backyard." The most significant issue is the high-level radioactive waste produced by nuclear reactors. According to the World Nuclear Association, about 9,000 metric tons of spent nuclear fuel goes into storage annually. Most of this fuel is saved for future reprocessing, but a permanent disposable option may be desirable. Releasing this waste from the end of a space elevator tether, far beyond GEO, would allow it to escape the Earth's gravity. Disposing of the full 9,000 MT/year would require 1.8 times the launch capacity of a single space elevator.

1.5.2 Earth sunshade

One ambitious idea to reduce global warming is by putting sunshades in space. University of Arizona astronomer Roger Angel has detailed this idea in the paper "Feasibility of cooling the Earth with a cloud of small spacecraft near L1" (Angel, 2006). His plan involves launching a constellation of trillions of small free-flying spacecraft a million miles above Earth into an orbit aligned with the sun, called the L-1 orbit. Angel proposes to design lightweight flyers made of transparent film pierced with small holes, two feet in diameter, 1/5000 of an inch thick, and weigh about a gram, the same as a large butterfly. The weight of all flyers would be 20 millions tons [Purang]. Launching this system would require 400 space elevators working for five years.

1.5.3 Space-based solar power

Another way to address global warming—and the growing power needs of human civilization—is the concept of Space Solar Power (SSP) satellites.

> *The National Space Society believes that one of the most important long-term solutions for meeting [our] energy needs is Space Solar Power (SSP), which gathers energy from sunlight in space and sends it to Earth. We believe that SSP can solve our energy and greenhouse gas emissions problems. Not just help, not just take a step in the right direction; solve. SSP can provide large quantities of energy to each and every person on Earth with very little environmental impact. [Mankins, 2011]*

Space Solar Power is gaining high-level attention because of its major attributes:

- A completely "green" and essentially unlimited energy source

- Capable of delivering energy to almost anywhere on the Earth
- Cost effective if launched with space elevators

For this global vision to be realized, the economic equation for space development, deployment and operations must change. Current business assessments of SSP do not make financial sense primarily due to the cost of launch to geosynchronous orbit. The existence of a space elevator transportation infrastructure will change the financial picture, making Space Solar Power practical.

The size and mass of a solar power satellite is explained by Dr. Chapman:

> *A single Space Solar Power satellite is expected to be above 3,000 MT, several kilometers across, and most likely be located in GEO, at 36,850km altitude, likely delivering between 1 to 10 GWe.* [Mankins, 2011, p. 31]

A satellite of 5,000 metric tons would take about one year to lift. (At 14 MT per lift and one lift per day, the throughput of a single elevator is 5,110 MT/year).

These three examples show that demand for space elevator lift operations could far outstrip the capacity that would be available in 2035. Such commercial mandates provide a strong incentive for further capacity development.

1.6   Spaceport locations—Where to park?

This discussion proposes that the space elevator community should not use the term "orbit" to define the location of a tether climber or its cargo, unless describing a trajectory after release from the tether. We should either specify altitudes or refer to the facilities along the tether that will be known as *spaceports*. Referring to these spaceports will help standardize names for altitudes for clarity of understanding.

History has shown that three classes of orbits (plus the transfer orbits that connect them) cover the vast majority of spacecraft missions: Low Earth Orbit (LEO), Geosynchronous Earth Orbit (GEO), and Middle Earth Orbit (MEO). Orbits around the Earth are designed to support specific missions. Common missions include communications satellites covering the Earth's surface, navigation satellites enabling essential location knowledge, and science missions looking both down and up. Each of these orbits has value and all are now crowded.

A real strength of a space elevator is that the tether climber can park at any altitude and conduct operations at that location. A logical extension of this concept is to position permanent spaceports at specific altitudes, attached to the tether in a way that allows tether climbers to bypass the stations as needed. Payloads can be lifted to these stations and removed from the tether for long-term operation. Table 1.1 shows potentially interesting spaceport locations.

Spaceports	Characteristic	Altitude (km)	Comment
Marine Node	Lower terminus	0	
Port of Call	Earth viewing	100–400	Good for hotels; 90% to 95% earth-normal gravity
LEO Port	Good view of LEO	1,500	Excellent location to study space debris
Science Port	Inside Radiation Belts	2,000	Good science location for many reasons
Release Port	Minimum height for satellite release	23,412	Results in elliptical orbit
Synchronous Port	Matches rotation of Earth	35,789	GEO node for satellite support
L-1 Port	Release and go to L-1	44,252	Rendezvous at L-1
Lunar Port	Release and go to the Moon	44,582	Depart for Lunar orbit or surface, or L-4/L-5
L-2 Port	Release and go to L-2	44,862	Rendezvous at L-2
Escape Port	Release to depart Earth	46,722	Escape to the other planets
Port Apex	Reach for outer planets	100,000	Location for Control of Tether Stability

Table 1-1. Where to Park

1.7 Phases of ascent and descent

Here are the phases of a typical ascent/descent mission for space elevator climbers.

Ascent phases

- Atmospheric Ascent (0 to 40 km): The first 40 km of the climb requires protection from atmospheric hazards, such as winds, lightning, rain, etc. We suggest providing this protection by means of a large, lightweight CNT protective box-like shroud as shown in Figure 1.2. This shroud will be described in detail in Section 3.1.
- Basic Ascent (40 to 35,786 km – GEO): The basic climber will use constant power to ascend using electrically driven wheels gripping the tether. The power is supplied by either a laser source on the ground or by ubiquitous solar power. As the climber rises, Earth's gravitational pull declines, and speed will increase. This portion of the trip should be accomplished within seven days.
- Beyond GEO Lift (35,786 to 100,000 km – GEO to Apex Anchor): Although this is a climbing operation, centrifugal force beyond GEO will tend to accelerate the climber away from the Earth. Accordingly, the emphasis during this portion of the trip will be on controlling the speed of ascent as it is being thrust outward. Solar power is needed only to operate the spacecraft itself. Rejecting the heat energy produced by the lifter's braking system will be a

major concern. Future designs of the beyond-GEO climbers might be significantly different from the ascent climbers fighting gravity during the climb from the surface to GEO.

Descent phases

- Descent from beyond GEO [100,000 to 35,786 km]: This phase will require climbing against the dominant force and need motors and energy [solar supplied] source for full trip. The design could be similar to the operational climber designed for the climb from 40 km to GEO.
- Descent from GEO (35,786 to 40 km): The same concerns apply here as described above for beyond-GEO lift. As the climber approaches the awaiting shroud at 40 km, the speed must be reduced to rendezvous safely.
- Atmospheric Descent (40 to 0 km): Once again, a shroud will surround the climber and its payloads to protect them from atmospheric hazards and return them to the surface of the ocean. This will require a shroud, motor, and gripping apparatus for braking of the total package.

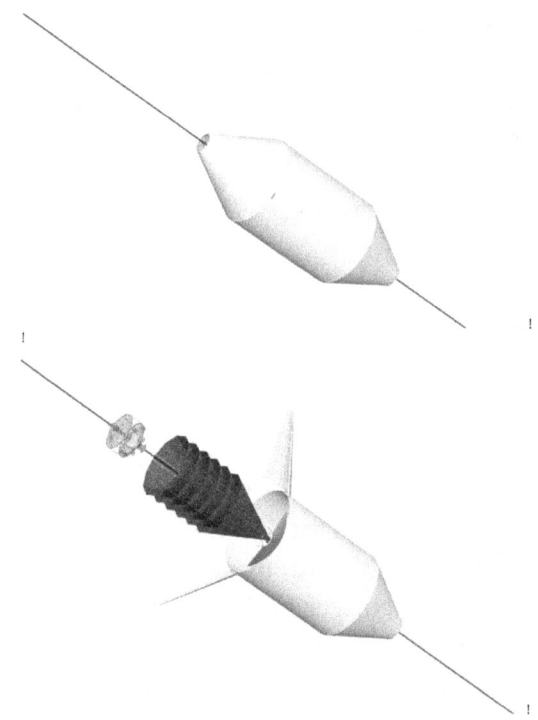

Figure 1-2. Climber protective shroud (chasedesignstudios.com)

1.8 Tether Climber Description

(Much of the following description is adapted from the IAA Cosmic Study [Swan, 2013].)

The word climber is used as the operative noun to denote the space system that is ascending or descending on the space elevator tether by its own means. The variety, in design, will be great. Operational climbers are defined as the commercial version of a spacecraft taking customer payloads to altitudes such as LEO, GEO and Solar System trajectories. It will also return objects to disposal orbits or to the earth's surface. The ascent requires power to climb while the descent from GEO and ascent past GEO requires braking as gravity or centrifugal forces dominate. The variety of operational climbers will surprise even early believers in a space elevator. There will be tether weavers, repairers, safety inspectors along with logistical trams, commercial climbers, human rated climbers, science climbers, hotels, and launch ports. An open standard will facilitate all manner of climbers to work on space elevators. The analogy would be the railroad's width of its rails. Anyone can put a train on the rails if they adopt the standards. A similar approach must be used to ensure compatibility between tethers and climbers. The complexity of the tether interface will drive the design of tether climbers and keep its maturity level between technological development and engineering applicability. Some of the baseline discussions have led to the following assumptions for this chapter and space elevator infrastructures:

- A capability of 31 metric tons for the space elevator.
- Each tether climber will be structured around a 20 MT gross weight at the Marine Node location with 6 MT allocated for the vehicle and 14 MT for the cargo. There could be a total of seven tether climbers on the space elevator at any one time, as shown in Table 1-2. The current estimate is one launch per day with trip time to GEO of approximately seven days.
- As tether climbers will be loaded and started at the Marine Node platform and protected through the atmosphere, the fragile solar arrays or laser receivers will be deployed on tether climbers at higher altitudes prior to initiating their climb.

Effective radius (km)	Effective g Factor	20 MT weight equivalent	6 MT weight equivalent
6,378 (at surface)	1.0	20 MT	6 MT
12,756	0.25	5.0	1.5
19,134	0.1	2.0	.6
26,600 (at GPS orbit)	0.043	0.8	.24
27,000	.041	0.8	.24
34,000	.014	0.3	0.08
42,160 (at GEO)	0.00	-	-
Total equivalent mass on tether of seven climbers		28.9	8.66

Table 1-2. Supporting seven climbers simultaneously

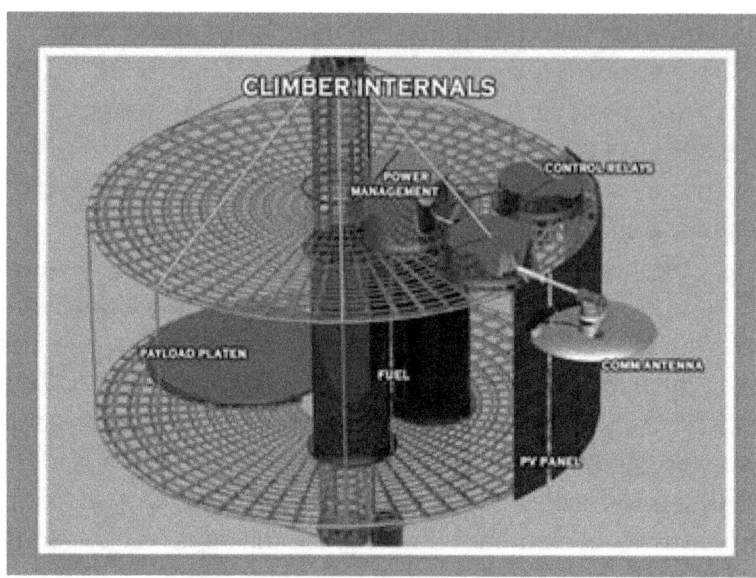

Figure 1-3. Tether climber components [chasedesignstudios.com]

1.9 Tether climber components

A tether climber is no more than a spacecraft with a special propulsion unit of electrically driven wheels instead of fuel-consuming engines (Figure 1-3). The components of a space elevator climber parallel a sophisticated spacecraft because of the similarities in requirements driven by the environment. However, a significant difference is the tremendously "softer" ride on an elevator vs. a rocket with its burning chemicals restricted by its rocket structure thrusting violently. An automated tether climber (with no crew) comprises the following major components:

- Structure (with cargo bays)
- Energy management (source, distribution, storage)
- Motor / Engine
- Tether interface equipment
- Drive apparatus (wheels)
- Communications payloads and antennas
- Environmental controls (heat, cold, vacuum, radiation, etc.)
- Attitude control (star sensors, GPS sensors, thrusters, spinning devices, controller, magnetic torque rods)
- Support equipment (robotic arm to load / off-load cargo, housekeeping, etc.)

The tether climber will be designed to operate above 40 km altitude with some type of external protection below that altitude for winds, lightning, rain, etc. This enables the designer to specify a tether climber that operates in the void of space and can be flexible in the design concept, as accelerations and shock loads will be minimal. This also allows designers to maximize their use of lightweight components, as the stresses are very low above the atmosphere. The tether climber will initiate its climb to mission altitude with cargo design criteria that are far less stringent than current launch vehicle design criteria.

1.10 Tether Climber Types

Multiple types of tether climbers will be required. Module-level standardization will be implemented where practical. Table 1-3 lists six types of tether climbers that span the known requirements.

Type - Mission	Designs	Power Option
Atmospheric Protection	Shroud to protect tether climbers within the atmosphere.	Power cable or laser
40 km-GEO	Principal operational climber to carry 14 MT of payload	Laser or solar
Beyond GEO	Could be new design or left-over principle climber design	Solar
Return from GEO	Probably operational climber returning to Earth without large solar arrays.	Solar
Repair	Variable design for different needs such as small patches or major segment replacement	Laser or solar
Construction	Early climber that attach tether material to original seed ribbon.	Laser or solar

Table 1-3. Types of Tether Climbers

Later chapters address specific design areas for tether climbers:

- Chapter 2 – Tether Climber Operations
  Describes operation of tether climbers from the surface of the ocean to the destination.
- Chapter 3 – Subsystems Design
  This chapter lays out the various components of the climber and discusses each. The major thrust is to identify needed design characteristics and risk mitigation plans to advance the level of technologies for the space elevator.
- Chapter 4 – Power Source Approach
  This chapter will address the current tradeoff study that is being conducted for two principal climber power options; laser beaming from the surface and solar power. (Previous studies have shown that internal power generation is impractical due to the size and weight of the necessary fuel.)
- Chapter 5 – Conclusions and Recommendations
  This chapter summarizes the study results.

- Appendix A - First 40 Kilometer Transit
  Discusses the various options to move the tether climber to 40 km in altitude so that it may start its climb to the stars.
- Appendix B – Tether Baseline
  Discusses tether design requirements to enable a space elevator infrastructure.
- Appendix C – Alternate Approach
  A variation in climbing technique is presented.
- Appendix D – The International Space Elevator Consortium

References

- Angel, R. (2006), "Feasibility of Cooling the Earth with a Cloud of Small Spacecraft near L1," Proceedings of the National Academy of Sciences, v 103, n46, 2006 November 14, 2006. Pp. 17184–17189. http://www.ncbi.nlm.nih.gov/pmc/articles/PMC1859907
- Bartoszek, Larry, "Getting the Mass of the First Construction Climber Under 900 Kg," ISEC Conference, Seattle, 2013.
- Edwards, B. C., and Westling, E. A., *The Space Elevator: A Revolutionary Earth-to-Space Transportation System*, BC Edwards, 2003
- Knapman, John, "Tether Climber at Constant Power," ISEC Conference, Seattle, 2013.
- Mankins, J. (2011), "Space Solar Power, The First International Assessment Of Space Solar Power: Opportunities, Issues And Potential Pathways Forward", IAA, October 2011.
- Mini-workshop Results, Tether Climber Brainstorming, ISEC Conference, Seattle, 2013.
- Penny, Robert "Skip", "Tether Climber Operational Phases," ISEC Conference, Seattle, 2013.
- Purang, Deepak (n.d.), "Space sunshade may one day reduce global warming." Editorial. http://www.streetdirectory.com/travel_guide/14921/gadgets/space_sunshade_may_one_day_reduce_global_warming.html
- Semon, Ted, "Hybrid Tether Climber," ISEC Conference, Seattle, 2013.
- Shelef, B. (2008c), "A Solar-Based Space Elevator Architecture," Spaceward Foundation, 2008. http://www.spaceward.org/elevator-library#SW
- Swan, P., Raitt, Swan, Penny, Knapman. *International Academy of Astronautics Study Report, Space Elevators: An Assessment of the Technological Feasibility and the Way Forward*, Virginia Edition Publishing Company, 2013.
- Lades, Martin, "Climber-Tether Interface for a Space Elevator, ISEC Conference, Seattle, 2013.
- http://en.wikipedia.org/wiki/Space_Elevator
- http://en.wikipedia.org/wiki/Wireless_power
- http://en.wikipedia.org/wiki/Radioactive_waste

Chapter 2   Tether Climber Operational Phases

The main mission of a tether climber is to deliver payloads to Geosynchronous Earth Orbit (GEO) and beyond. This chapter will describe the activities of the climber as managed from the Climber Operations Center in the Headquarters and Primary Operations Center (HQ&POC) as they ascend or descend. It will cover the operational phases of a climber from its delivery to the Base Support Station at the Marine Node, and from there to the GEO node and back. This chapter will delve deeper into operational aspects of climbers first introduced in ISEC Report 2012-1, Space Elevator Concept of Operations [Penny, 2013]. Figure 2-1 shows an operations perspective of the facilities involved in space elevator operations.

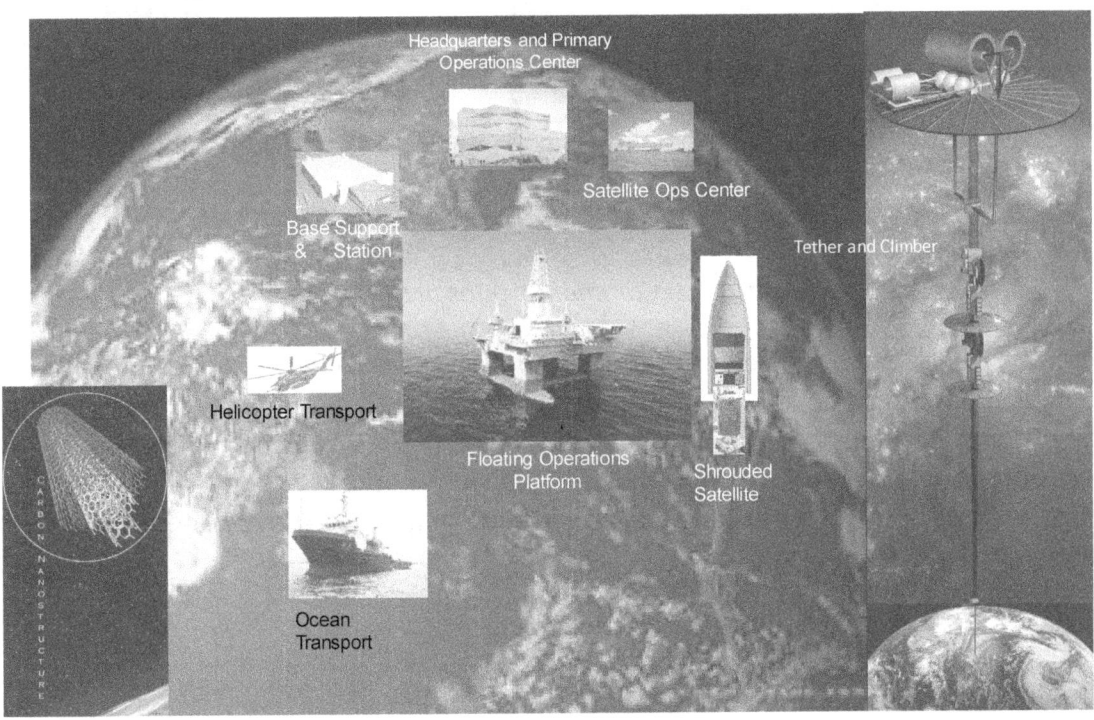

Figure 2-1. Space elevator facilities

Communications across the space elevator architecture will have a consistent theme: every data transmission goes through the GEO node Communications Infrastructure to the HQ&POC. As such, climbers will be in constant contact with the Climbers Operations Center at headquarters via the communications packages at the Marine Node, on each climber, at GEO node, at the Apex Anchor and beyond GEO. Figure 2-2 shows the communications architecture for climber operations.

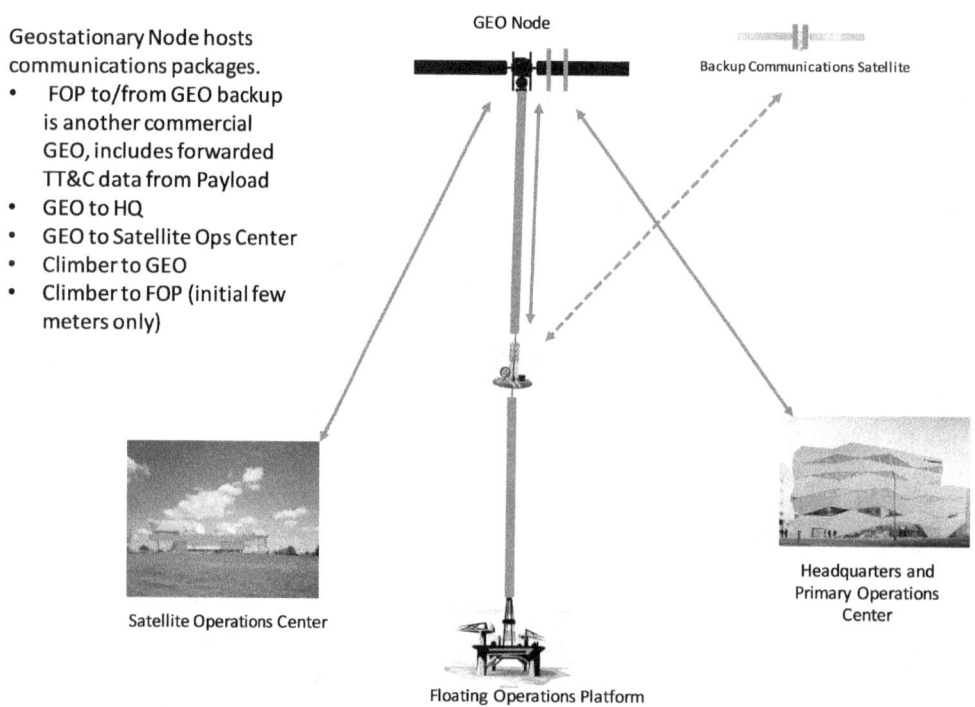

Figure 2-2. Communications architecture

In addition, the whole study baselined a concept for two space elevators working together as a pair. The first one would be for "up" traffic principally while the second space elevator would facilitate "down" traffic as needed. This enables the commercial approach to emphasize the strength of the space elevator – moving mass off-planet. Another key point would be that the second space elevator would be a redundant capability so necessary in commercial businesses in case of emergencies or catastrophes.

2.1   Climber types and descriptions

There will be at least three types of operational climbers: one for delivery of cargo (all altitudes up to GEO), one for delivery above GEO, and one for tether repair. Cargo will consist of satellites destined for Low Earth Orbit (LEO), Medium Earth Orbit (MEO), Geosynchronous Earth Orbit (GEO), interplanetary trajectories, and the Apex Anchor counterweight. Additionally, cargo will include supplies for GEO node operations (including fuel for space tugs). The repair climber will be equipped with sensors to detect holes and tears in the tether and equipment to conduct repairs. The climber used to deliver payloads above GEO will need only a small motor and solar arrays for the return journey to GEO.

2.2   Phases of Operations

To understand the design and use of operational tether climbers, a concept of operations must be presented. To address climber operations, we suggest there could be operations phases as defined in the following paragraphs. The vision of an architecture is given in Figure 2-3, Space Elevator Nodes.

2.2.1 Preparation for climb

This is the phase that starts with transport from factory to Floating Operations Platform (FOP) at the Marine Node and placement in storage. Preparation continues with removal from storage and configuration for attachment and climb. This phase includes such things as installing the solar array package (to be deployed at 40 kilometers), loading of consumables such as fuel for cold gas thrusters, charging the battery, installing parachutes, installing the protective equipment for the atmospheric phase, and so on. The cargo placement is designed to ensure center of gravity (C_g) and to keep the z-axis of the climber aligned with the tether. Local weather is monitored as well as forecasts at altitude.

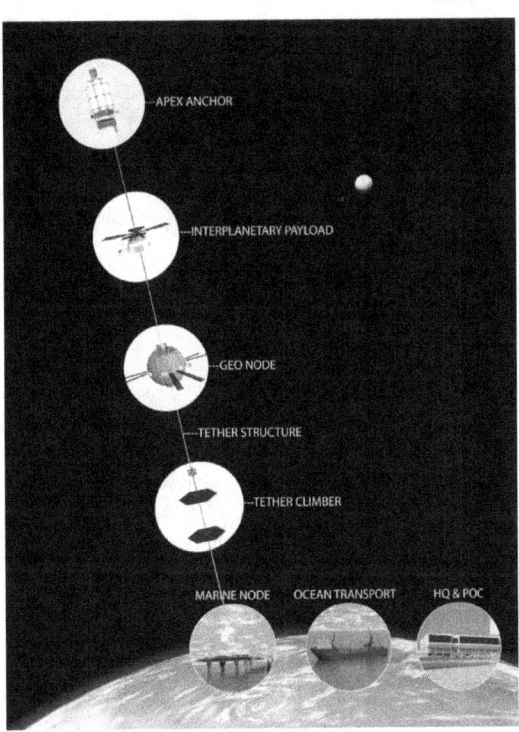

Figure 2-3. Space Elevator Nodes [chasedesignstudios.com]

2.2.2 Attach

This is the phase where the climber, on its dolly, is rolled to the proximity of the tether. The gripping mechanism is opened and the climber is positioned to enable the gripping mechanism to clamp on the tether. Cargo is loaded per the C_g management plan. The climber and payloads are then placed inside the atmospheric protective shroud. A power cord is installed and systems checks are performed including communications with the climber Ops Center. Figure 2-4 below depicts the mating activity. This phase includes loading of the customer's payload(s) when not done before mating.

Figure 2-4. Mating climber to tether (chasedesignstudios.com)

2.2.3 Atmospheric lift (under 40 km)

This is the operational phase for the initial 40 km climb to an altitude above the atmosphere where solar arrays or laser receivers can be deployed safely. The protective shroud uses a low gear, as the weight of the climber, its payload and the shroud is the greatest at sea level. The rate of climb has to be a compromise between the engineering capabilities of the climber and the need to minimize the total load on the tether. A speed of 40 kph seems reasonable. Start time of the climb will be chosen so that the climber arrives at 40 km at the beginning of the solar day. Starting as late as possible before dawn allows the other climbers on the tether to be as high as possible, where the gravity is low. When above the atmosphere, the climber (with payloads) ascends out of the shroud and initiates its climb. The shroud returns to the surface.

2.2.4 Heavy gear, high gravity

This phase starts above the atmospheric phase (40 km) and continues to an altitude where a higher gear can be employed. Changing to the next gear will likely be done while stopped since shifting on the fly is more complex. A constant power approach will ensure lighter designs with varying speeds (to be discussed in the next chapter).

2.2.5 High speed

This is the phase where speeds increase due to higher gears being used with less gravity effect and constant power. Climbers will also operate at high speeds on descent.

2.2.6 Detach at GEO node

This is the phase where GEO and beyond-GEO payloads are off-loaded at the GEO node operations center. GEO satellites will propel themselves to their assigned slot (longitude). LEO payloads will have been dropped off earlier. Other non-satellite cargo is also off-loaded. This is all done robotically.

2.2.7 Coast to beyond GEO

This is the phase where payloads "coast" to higher altitudes for release into non-Earth orbits or just to the Apex Anchor. Various release altitudes initiate flights to the Moon (or L1/L2), Mars and beyond. This operational phase also includes payloads that continue on to the Apex Anchor to become part of the counter-weight.

2.2.8 Load cargo

This is the phase where cargo is loaded into the descending tether climber for drop off in disposal orbit or return to the earth's surface. There will be a constant monitoring of the center of mass for the package and climber. This could be accomplished at the GEO node on either the up or down space elevator.

2.2.9 Transfer to descending tether

This is the phase where, once detached from the ascending tether, the climber is transported via space going tug to the GEO node of the descending tether.

2.2.10 Attach to descending tether

This is the phase that mimics the actions of attaching at the Marine Node. Cargo can include retrieved satellites that can be placed in re-entry orbits or simply returned.

2.2.11 Descent to 40 kilometers

This is the phase where the climber, not really "climbing" now, descends down the tether. Brakes will help manage speed.

2.2.12 Atmospheric re-entry

This is the phase that reverses the ascent through the atmosphere. The descending climber slows down and matches the altitude with the waiting protective shroud and then enters the shroud. The protective shroud then descends through the atmosphere and softly approaches, and docks with, the Marine Node.

2.2.13 Detach

This includes unloading of cargo such as retrieved satellites. This phase is the reverse of the Attach phase. All payloads would be off-loaded, as well as the tether climber being released from the atmospheric protective shroud.

2.2.14 Transfer to ascending tether.

This is the phase where the climber is transported by sea to an ascending tether at another Marine Node.

2.3 Climber Missions

To understand the design of tether climbers, the projected missions must be discussed. The following sets the stage for operational tether climber types.

2.3.1 GEO Delivery and Pickup

GEO delivery of satellites will be the primary commercial activity of the Space Elevator. GEO satellites will be less expensive to build, as they don't have to survive a rocket launch environment; and, they don't need the redundancy associated with 15-year mission life. Delivery will include satellites destined for higher or interplanetary orbits as well. Finally, delivery will resupply the GEO node including fuel for tugs or other orbit delivery systems. GEO pickup will include spent satellites that would be returned for refurbishment or scientific study. Other spent satellites would be released in rapid decay orbits at suitable altitudes below GEO.

2.3.2 LEO Delivery

When a cost effective space tug has been developed, LEO satellites will be dropped off below the GEO node and placed in the desired orbit (apogee, perigee, and inclination) by a space tug. LEO satellites will enjoy the same less expensive costs as their GEO counterparts for the same reasons. As the space tug technology matures, it may also be used to place spent LEO satellites and space debris into rapid decaying orbits for disposal.

2.3.3 Beyond GEO Delivery

There will be many missions for the space elevator because of its ability to release spacecraft with significant energy to reach solar system targets. At various altitudes, the energy is sufficient to reach L-1, L-2, the Moon, Mars and beyond. There will be a climber that picks up spacecraft at the GEO node and takes them to the proper drop off location and then returns to the GEO node.

2.3.4 Where to Park?

This mission could also be to place an object (scientific payload, weather watching satellite, etc.) at an altitude and allow it to just stay on the tether and operate. As such, a new idea is enabled with space elevators: The ability to park. This would require a good parking brake and coordination with all up and down climbers.

2.3.5 Return to Earth's surface

Most often the climber will return to the earth's surface using a combination of its own power and gravity with braking.

2.3.6 Tether Repair

Tether repair will routinely occur on a scheduled basis or, during any emergency, as needed. All tether climbers will have sensors that detect wear and tear on a tether and enable climber operations personnel to schedule repairs. Tether repair climbers move much slower than routine climbers; so, deliveries may be reduced during repair operations. Repair climbers will be able to conduct repair operations while ascending and/or descending.

2.4 Climber Operations Center

The Climber Operations Center will be located at the Headquarters (HQ) and Primary Operations Center (POC). The POC will have multiple big screen displays to provide situational awareness of the Space Elevator System of Systems. There will be a "top level" display for the system with drop-down windows for each of the major components. For climber operations, there will be a display showing all tethers (ascending, descending, under construction) and all climbers on the tethers. Each climber will have a color background indicating operational state: green for fully operational; yellow for partially operational; and red for not operational. The display will indicate missions and phases for each climber. Similar treatment will be applied to the GEO node, the tether terminus on a Floating Operations Platform (FOP), and at the Apex Anchor.

Each of the climbers will be in constant contact with the POC sending its health and status telemetry and receiving commands from operators, loading and unloading of satellites at the GEO node, releasing satellites below GEO, and other missions.

Figure 2-5. Notional Operations Center

2.5 Life Cycle of a Climber

The authors choose to use the standard satellite delivery climber as our subject, choosing the GEO delivery mission to chronicle, and use solar power as the source of energy for a life cycle example. This chapter talks about the ascending and descending tethers. In addition, the second tether becomes the essential redundant tether for potential failure of the principal tether. The initial concept is that both tethers will be used for commercial up traffic (major source of revenue) and when needed, the second tether would convert to a downward path. This is vitally important as the descending tether frees up the ascending tether to earn revenue from ascending climbers.

- The climber is delivered to the Base Support Station (BSS) at the Headquarters and Primary Operations Center. It arrives in the environmental protective cover needed for its transport to the Floating Operations Platform (FOP) at the Marine Node.
- The climber is in its protective case, has plugins for power (not the high power needed for the drive motors), and health/status monitoring and control. The climber is checked to ensure no damage has occurred during transit, from the factory to the BSS. It is anticipated that virtually every time the climber will be deemed ready for shipment to the Ocean Going Vehicle (OGV). It is placed in a powered down mode and loaded on the OGV with supplies and payloads destined for the Marine Node. The climber remains powered down during the five-day trip to the Marine Node. Upon arrival at the Marine Node, the climber is off loaded and moved onto its dolly for storage; or, it may be moved directly to the proximity of the tether. Upon arriving in the mating area, the climber is powered on and its health/status

checked. This is done through an umbilical that provides information to the local operations center and back to the HQ/POC.
- Usually, the climber will not need maintenance. If it does, it can receive maintenance on the spot or rolled back to a maintenance area. Very rarely would it need to be sent back to the BSS for shipment to the manufacturer.
- The above steps occur the first time a tether climber is deployed. Since the climbers are reusable, the remaining steps are repeated for every climb.
- The protective cover is removed and payloads and counterweights are positioned in the climber. A center of gravity management plan will have been generated to dictate it's positioning.
- The climber is then installed in the protective shroud with its own motor and tether gripper.
- Weather will be checked using data from the FOP as well as national and international agencies at the HQ/POC. Mating with the tether will only be scheduled for times when the weather allows.
- The climber, inside its atmospheric protective shroud, is next "mated" to the tether. This may include several attachment points and movement up the tether to complete all the attachments. Thus, connection to the power cord (sufficient energy to provide power to the drive motors) will be accomplished when needed to permit the climber to rise high enough to complete all attachments.
- Health and status monitoring will continue through an umbilical until the climber is ready to begin ascent. Communications through the GEO Comm's Node back to the Climber Operations Center will be established and the umbilical will be detached.
- The climber now begins its climb in the atmospheric phase which overlaps with the Heavy Gear, High Gravity phase. Control is transferred from the FOP to the HQ/POC. The climber is powered by the heavy-duty extension cord being reeled out from the FOP. A mechanism will probably be needed to prevent the cord from coming in contact with the tether. FOP personnel monitor the power cord and are in contact with the POC.
- Start times will be selected so as to reach 40 km at the beginning of the solar day. As an example, if the climber were capable of 40 kph, the trip would take about an hour. To arrive at 0530, it would start at 0430.
- All during this phase the health and status telemetry is being monitored by the POC. The climber will be highly autonomous; but, POC operators will be able to send commands to the climber while receiving telemetry.
- Upon reaching 40 km, the atmospheric protection shroud will stop and set its parking brake. The tether climber will rise out of its shroud on stored power. It will then deploy its energy arrays and begin using them as source to power the drive motors. A different, higher gear may be selected for the continuation of the climb. The climber will resume climbing at increased speeds in the higher gear. It will stop to change gears as needed and stop for solar night during the rest of the climb to GEO. The stops will be dual purpose. The climber uses a battery for life sustainment during the night.
- The climber protective cover shroud, power cord, and initial lift apparatus would then be lowered back to the FOP.

- The climber will continue climbing (and changing gears) during daylight hours until it reaches the GEO node (decelerating to a safe speed for docking with the GEO node).
- The climber will be de-mated from the tether and its cargo moved to a storage area at the GEO node. Its solar arrays will be reeled in. GEO satellites will either be released and propel themselves to their orbital slot or they may be mated to a space tug for transfer. Figure 2-6 shows some NASA concepts for those tugs.

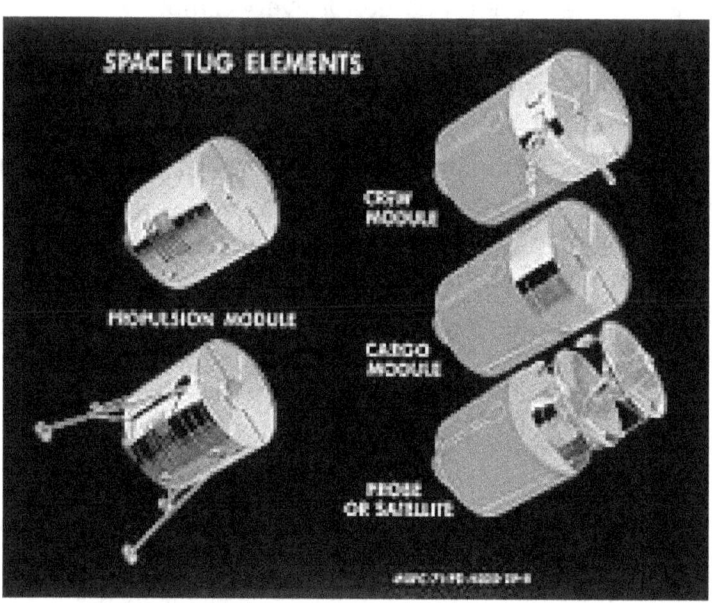

Figure 2-6. Tug Concepts from NASA Images (MSFC-71-PD-4000-29-B)

- The key attribute of these tugs will be their ability to mate with payload GEO satellites, deliver to orbital slot, de-mate with satellites and return to the GEO node. A design goal will be to have the tug capable of rendezvous and capture of non-cooperative GEO satellites for transport to either the ascending node tether (to become counterweight at the Apex Anchor) or the descending tether node (for drop off at a disposal altitude).
- Space tugs will also be used for placing LEO satellites in their desired orbit and returning to a GEO node. The same capability for capture of non-cooperative debris will be desirable for mitigation purposes.
- Satellites destined for disposal orbits or returned to the Earth's surface will be loaded at the descending tether's GEO node. The same center of gravity planning will have occurred to determine placement in the climber. Solar arrays will be re-deployed as the power needs are greatly reduced.
- The climber's motors will get the climber headed back towards the earth's surface with large contributions by gravity. The cycle of stops for nightfall and changing gears will not be needed. Speed control with braking and heating will be the primary concern. The excess energy could be stored or transmitted for use elsewhere.
- At 40 km, the climber would re-enter the protective shroud for transfer through the atmosphere. The climber may pause to avoid unfavorable weather in the atmosphere;

however, the duration of the last 40 km should be much less than the climb in the first 40 km.
- The protective shroud, with the climber, will continue its descent slowing to a safe speed for docking with the FOP. Upon arrival, the power and umbilical cords will be connected. Satellites will be unloaded and taken to a processing area where they will be packaged for the trip to the HQ/POC on the OGV.
- The climber will then be "de-mated" from the tether and placed on its dolly. It will be wheeled to a temporary storage area and installed in a protective cover. When the OGV arrives, the climber will be loaded and transported to an ascending tether FOP. It will be wheeled to a refurbishment area where it will be inspected and repaired as necessary to return it to fully capable status for its next climb.

2.6 Operations growth potential

This look at potential approaches to space elevator operations, as described in this chapter, enables the infrastructure to expand logically. Once that pair of space elevators has been operationalized, more pairs can be sold around the world. The beauty of the operations concepts presented in this chapter is that they are expandable and redundant, enabling multiple pairs of space elevators.

2.7 References

Penny, Robert. Swan, Peter, & Cathy Swan, "Space Elevator Concept of Operations," ISEC Position Paper #2012-1, International Space Elevator Consortium, Fall, 2013.

Chapter 3   Sub-System Description

The systems engineering approach for tether climbers spans all the engineering and scientific disciplines. The good news is that a tether climber is no more than a traditional spacecraft with a unique approach to lift. The motor and tether gripper apparatus will be unique to space elevator systems; but, the other elements all have historic models. A significant difference, and benefit, is the tremendously "softer" ride on an elevator vs. a rocket with its burning chemicals restricted by its nozzle structure thrusting violently. In addition, the volumetric requirements are significantly sized to allow large structures to be deployed without forcing them into a faring. This study report incorporates much information from, and is an expansion of the work accomplished for, the IAA Cosmic Study entitled: Space Elevators: An Assessment of the Technological Feasibility and Way Forward [Swan, 2013].

3.1   Climber Types

During the discussion on tether climbers in chapter one, various types of climbers were explained. A variety of climbers will represent the spectrum of future needs. However, this report will assume the basic operational space elevator tether climber with a mass of 20 MT. This will ensure the chapter flows consistently. There will be some discussions on the construction climber and an initial discussion on the atmospheric shroud to help the reader understand major factors in the design. As a reminder, the types of climbers are:

Type - Mission	Designs	Power Option
Atmospheric Protection	Needs protection from atmosphere and power from below.	Power cable or laser
40 km-GEO	Principal operational climber to carry 14 MT of payload	Laser or solar
Beyond GEO	Could be new design or existing principal climber design	Solar
Return from GEO	Probably operational climber returning to Earth without large solar arrays.	Solar
Repair	Variable design for different needs such as small patches or major segment replacement	Laser or solar
Construction	Early climber that adds tether material to original tether seed ribbon.	Laser or solar

Table 3-1. Climber Types

3.2 Atmospheric Protective Shroud

While traveling through the atmosphere, tether climbers and payloads are vulnerable to atmospheric threats such as high winds, lightning, and rain. The tether climber could be designed to enclose the cargo within a permanent protective structure, as current rocket-based launch systems do. However, this structure will not be needed once the climber leaves the atmosphere. Since exo-atmospheric travel represents the vast majority of the lift to orbital release altitude or GEO, it is inefficient to carry atmospheric protection over the whole distance. Instead, we propose using a separate box-like protective shroud to enclose the tether climber only while it is in the atmosphere below about 40 km altitude, at which point atmospheric threats essentially cease to exist. This strategy of localized protection significantly reduces the mass of the operational climber. Figure 3-1 shows a climber ascending out of its shroud at the 40 km point.

Figure 3-1. Tether climber emerging from shroud (Chasedesignstudios.com)

The box will have its own climber motor and tether interface apparatus so that it can climb to 40 km and return. The power to the shroud and the internally stored operational climber is assumed to be provided via a long, lightweight power cord. During the mission to 40 km, the box will provide communications between the FOP and the climber and its payload. Although such a shroud has not been designed in detail, the following assumptions should establish the concept and support the strengths of the idea.

- Motor and attachment apparatus similar to operational climber
- Structure is composed of a platform for operational climber to sit upon and a series of struts to position the shroud
- Hatch at the top for climber exit and entry while on the tether
- The shield, structure, and power cable will likely be made with CNT-based composite materials

This protective shroud system will indeed consume some of the mass allocation for the total carrying capacity of the tether. For a starting point, the mass of the shroud system is assumed to be somewhere around ten metric tons. This would enable the tether carrying capacity to be similar to the estimates for previous studies with the one exception: the effective throughput would probably be six tether climbers going to GEO per week vs. seven. This approach will need far more analysis to refine the throughput and impact of this design to the total space elevator infrastructure. However, the separation of design criteria into two distinct sets (inside vs. above the atmosphere) enables optimization for design – an old trick by designers when requirements can be divided into distinct and separate categories.

3.3 Operational Tether Climber Design Basics

The tether climber is composed of the following major components:

- Structure (with cargo bays)
- Energy management (source, distribution, storage)
- Motor
- Tether interface equipment
- Drive apparatus (wheels)
- Communications payloads and antennas
- Environmental controls (heating, cooling, vacuum, radiation, etc.)
- Attitude control (star sensors, GPS sensors, thrusters, spinning devices, controller, magnetic torque rods)
- Support equipment (robotic arm to load / off-load cargo, housekeeping, etc.)

This chapter is focused upon early operational robotic climbers carrying customer payloads to GEO and beyond. Another major section discusses the issue of friction between the wheels and the tether as well as a quick look at a wheel design. In addition, a portion of this chapter will propose a size and design for the construction climber which "builds up" the total operational tether from an initial seed tether at deployment to full capacity by adding tether mass. The last portion of this chapter opens the general discussion about tether climbers with further thoughts gained through brainstorming sessions at the 2013 International Space Elevator Conference in Seattle.

3.3.1 Customer Needs

The needs of our customers must be addressed early in the design process to ensure proper tradeoffs in all engineering choices. The basic concept that will be leveraged in this report is based upon the 2013 International Academy of Astronautics study report [Swan, 2013]. The first step in a design process is to identify the criteria. This is usually consolidated in the Systems Design Requirements Document early in the design-development process. A preliminary set of requirements for operational space elevator tether climbers is shown in Table 3-2.

Requirement for Climber	Segment	Easy Hard
Spacious for customer's 14 MT payload	S	easy
Balanced for center of mass	S	routine
Seven days to GEO	M	hard
Sufficient power for climber	E	medium
Pointing solar arrays/laser receivers	A	medium
Power storage and distribution	E	easy
Attitude control of climber	A	medium
Communicate with operations and customer	C	easy
Compatible with Marine Node	S	medium
Compatible with GEO Node	S	medium
Survivable in space environment	Ev	medium
Sufficient coefficient of friction	S	medium
Maintain contact with tether	M	medium
Tether designed for stress/strain of climber	M	medium

Table 3-2. Requirements Set for Climber
(Segment key: S-structure, A-attitude, E-energy,
C-communications, M-motor, Ev-environmental)

When one uses this set of criteria and the design from the Academy study, the baseline results show there could be space elevators based upon:

- One week trips to GEO
- 20 MT climbers with 14 MT payloads
- Seven climbers on the tether simultaneously
- External power sources
- Atmospheric protective shroud for lowest 40 km of travel
- Space elevator infrastructure that should be profitable to both customers and owners

3.3.2 Simultaneous Tether Climbers

The complexity of the tether interface will drive the design of tether climbers and currently has a maturity level between technological development and engineering applicability. Some of the baseline discussions have led to the following assumptions for this chapter and the space elevator infrastructure when based upon a solar energy source (similar assumptions for laser energy sources lead to similar discussions):

- A load capacity of 31 metric tons (MT) for the space elevator tether. This means that the tether can suspend a single 31 MT mass at the Marine Node with an adequate safety margin.
- Each tether climber will have a 20 MT gross mass with 6 MT allocated for the vehicle and 14 MT allocated to the cargo.
- As Table 1-2 showed, spacing multiple tether climbers 6,378 km apart allows up to seven climbers to share one tether within the 31 MT load limit. This is possible because the effect of gravity is significantly reduced at higher altitudes, so the higher tethers represent significantly less load. [Swan, 2013]
- Thus, the current estimate involves one launch per day with trip time to GEO of approximately seven days (assumes an average velocity of 215km/hr or 60m/sec).

As the tether climber will be loaded and started at the Marine Node platform and protected through the atmosphere, fragile solar arrays or laser receivers will be deployed from tether climbers at altitude prior to initiating the climb above 40 km. For the solar option, the early morning sun will start the climb for the vehicle, which will go against gravity until night occurs. During this eclipse, the vehicle will park on the tether and conserve heat and energy until the next morning (significantly shorter than 12 hours as it is more than 3,000km away from the surface of the Earth). Each day's climb will have longer and longer "daylight" periods until it reaches continuous solar energy (except for twice a year at spring and fall equinoxes when there will be short daily eclipses all the way to GEO and beyond).

When the power source for the tether climber is laser, the operational procedures will be less "daylight" dependent and more distance dependent. The first forty kilometers will be provided inside a protective box regardless of the power source to be used at higher altitudes. Once released above the atmosphere, the laser energy based tether climber will initiate its lift off and continue to rise. A strength of this concept is that while the collimated beam energy falls off as a function of distance, the gravitational force falls off even more rapidly.

3.3.3 Constant Power Baseline

Recent work, including the Cosmic Study for the International Academy of Astronautics [Swan, 2013], has suggested that tether climbers will ascend at an average speed of 60 m/s (216 km/hr). They start climbing at dawn and continue until they enter the Earth's shadow. At ground level, a 20 MT climber needs nearly 12 megawatts (MW) of power to achieve that speed. As it ascends, it experiences a lower gravitational force towards the Earth and the power consumption drops. At 4,470 km altitude, the power required is only 4 MW. As the climber approaches GEO, the lifting power requirement drops to zero. This means that for a constant-speed drive, roughly two-thirds of the cost and weight of a 12 MW drive system is used for only about one sixth of the ascent.

Mr. Ben Shelef has pointed out that it is also possible for climbers to ascend at constant power [Shelef, 2008c]. This means that tether climbers will make slower progress at lower altitudes, where it puts a greater load on the tether. As the climber ascends, its effective weight diminishes because of reduced gravity, so smaller motors and power supplies are needed to maintain a given speed.

The drawback of the constant-power approach is that climbers travel more slowly up the lower parts of the tether. This causes an extra load on the tether as climbers are bunched together at lower elevations where their weight is greater. In the previously referenced Academy study, it is shown that the tether climbers reach an altitude of 4,070 km in the first day at a constant 60 m/s (allowing for a wait of 5.4 hours during the night).

On the other hand, at a constant 4 MW, a 20 MT climber starts at a speed of only 21 m/s; it reaches 32 m/s at the end of the first day, at which point it has reached an altitude of only 1,580 km, where its weight is 12.7 tons. However, it can make up for this slow progress later on and still reach geosynchronous altitude (GEO) within seven days even if the top speed is limited to 83 m/s (300 km/hr). Figure 3-2 shows both arrangements with the weights of the tether climbers at various altitudes. We assume that the climber commences its ascent from 40 km where there is already a small weight saving, so that a 20-ton climber weighs 19.7 tons, as shown. The vertical dots are locations after each 24 hours of travel.

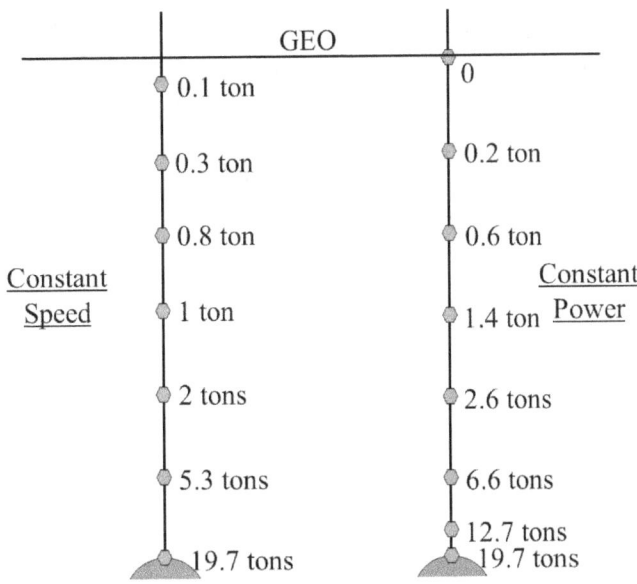

Figure 3-2. Disposition of tether climbers under constant speed (left) or constant power with speed limit (right)

The figure on the right, of a 4 MW constant power option, was chosen as a recommendation. It is, however, possible to trade off between the power rating, maximum speed and time to GEO. The following table is taken at an equinox, which gives the slowest times of ascent.

	Max speed 250 km/hr	Max speed 300 km/hr	Max speed 1000 km/hr
Power (MW)	Days to GEO	Days to GEO	Days to GEO
2	10½	10	8
3	9	8	5½
4	8	7	4
5	7½	6½	3½
11.8	7	6	2

Table 3-3. Trade-Space: Time to GEO

A slightly lower power rating would also be acceptable, especially if it can be combined with a greater maximum speed at the higher elevations. For example, 3.5 MW power and a maximum speed of 400 km/hr leads to a time to GEO of 6 days.

A factor of three savings in the power requirements for a tether climber means that the projected shapes of solar panels required can be reduced from 200,000 m² to 66,000 m². This reduction in area allows the mass density of the array to increase from 15 gm/m² to 45 gm/m² while maintaining the same total array weight assumed in the 2013 IAA Cosmic Study. Alternatively, we can relax the requirement for photovoltaic (PV) cells to have 70% efficiency. As at present 10 to 18% is typical, this will enable usable PV cells much sooner. We can now update the table in the IAA Cosmic Study. Here the "current array" is a prototype deployed in a German experiment. The future projections are based on a USAF study [USAF, 2012].

---

[1] An AF study showed that for every 1% increase in solar cell energy generation efficiency translates to a 3.5% increase in power [or decrease in mass]. [USAF 2012]

	Current array	Scaled up for 20-ton tether climber	Future array, medium efficiency	Future array, high efficiency
PV efficiency	10 %	10 %	40 %	70 %
Output (kW)	50	4,000	4,000	4,000
Improvement[1]	Current	Current	30% efficiency or 105% power	60% efficiency or 210% power
Mass required (kg)	32	2,500	1,220	810
Area required (m²)	400	200,000	100,000	66,000
Size of square array	20 m x 20 m	447 m x 447 m	316 m x 316 m	257 m x 257 m

Table 3-4. The effects of solar array efficiency

A budget of 6 tons has been allowed for the tether climber structure and mechanism. Clearly, a 4 MW motor is much easier to build than one of 12 MW, and it can be much smaller and lighter. The choice to go to constant power seems optimum at this time. The choice of operational power level for solar energy climbers will depend upon technological advances in photovoltaic cells, such as efficiency, mass density, and area needed.

3.4　Tether Climber Sub-Systems

The breakout of subsystems for a tether climber is similar to any spacecraft system with one major exception – the substitution of a motor and tether coupling mechanism instead of a rocket motor. The first part of this section will describe structural mass reductions enabled by the use of carbon nanotubes. This section will then break out to cover other spacecraft subsystems. The following design is for an operational robotic climber designed for a 4 MW constant solar power approach. (Similar designs result from the choice of laser power sources.)

3.4.1　Mass and Cost Savings Possible With Carbon Nanotubes

Future spacecraft will be lighter and more efficient because of breakthroughs in materials science, especially the introduction of composite materials using carbon nanotubes (CNT).

Table 3-5 shows actual mass measurements for two existing satellites, estimates of the weight reductions possible in these satellites with the use of CNT materials for strength, stiffness, electrical conductivity, electrical insulation, heat transfer and heat isolation, and mass estimates for corresponding elements of the proposed tether climber.

	FLTSATCOM 6		DSP 15		Tether climber
	Original Mass	With CNTs	Original Mass	With CNTs	With CNTs
Payload	226	57	563	141	555
Structure	168	42	408	102	1,500
Thermal	17	4	42	11	120
Power	340	85	825	206	2,340
TT&C	26	7	63	16	180
ADCS	52	13	127	32	360
Propulsion	35	9	85	21	740
Dry mass	871	218	2,115	529	5,795
Propellant	81	20	162	41	205
Wet mass	952	238	2,277	569	6,000

Table 3-5. Mass reductions enabled by use of carbon nanotube materials for two existing satellites plus the proposed tether climber. (All masses in kg.) [SMAD. pg. 894]

As the table shows, the surprising result is that CNTs can reduce spacecraft mass by up to 75%. Cost savings can be similar, since to a first order, production cost and mass are directly related. These savings easily lead to more spacecraft and more missions accomplished in space for a given budget and launch capacity.

3.4.2 Structure

Figure 3-3 shows a notional layout for the structural elements of a tether climber. This structure is about 20 meters in diameter and 15 meters in height, thus providing a cargo volume that is much larger than is practical with rocket launch systems. The extra volume allows low-density payloads such as lightweight trusses and large antennas—payloads that are further enabled by the low-stress lift conditions provided by space elevators.

The tether passes through the central tunnel, where it interfaces with the motor and wheels. The rest of the structure is balanced to minimize stresses on the tether. The estimated mass for a climber of this size would be around 6 MT.

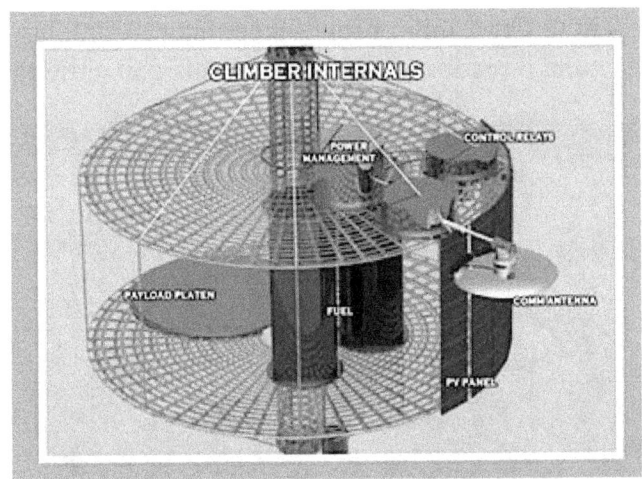

Figure 3-3. Tether climber structure [chasedesignstudios.com]

Figure 3-4 shows a tether climber with its solar arrays deployed for energy collection, with the majority of the cells being supported below the climber structure. The design will ensure that the pointing towards the sun is automatic as the angle changes over the daylight periods. The concept is a series of strings, as guides, would ensure all solar arrays pointed in the same direction with the rotation around the tether achieved for the afternoon sessions.

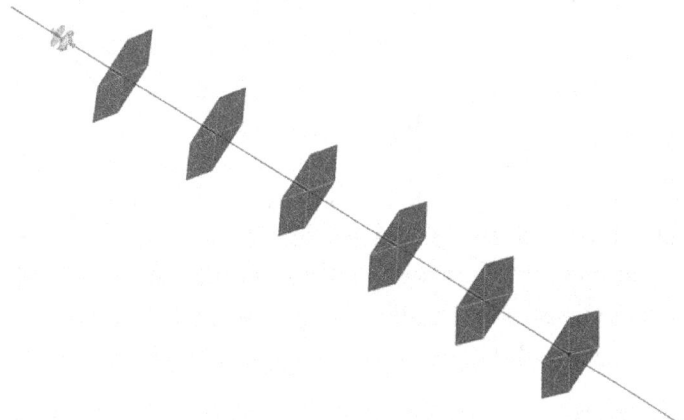

Figure 3-4. Tether climber solar arrays [chasedesignstudios.com]

Figure 3-5 shows a different tether climber proposal configured with laser energy receivers. The pointing is simpler as the nadir direction is essentially stable along the tether.

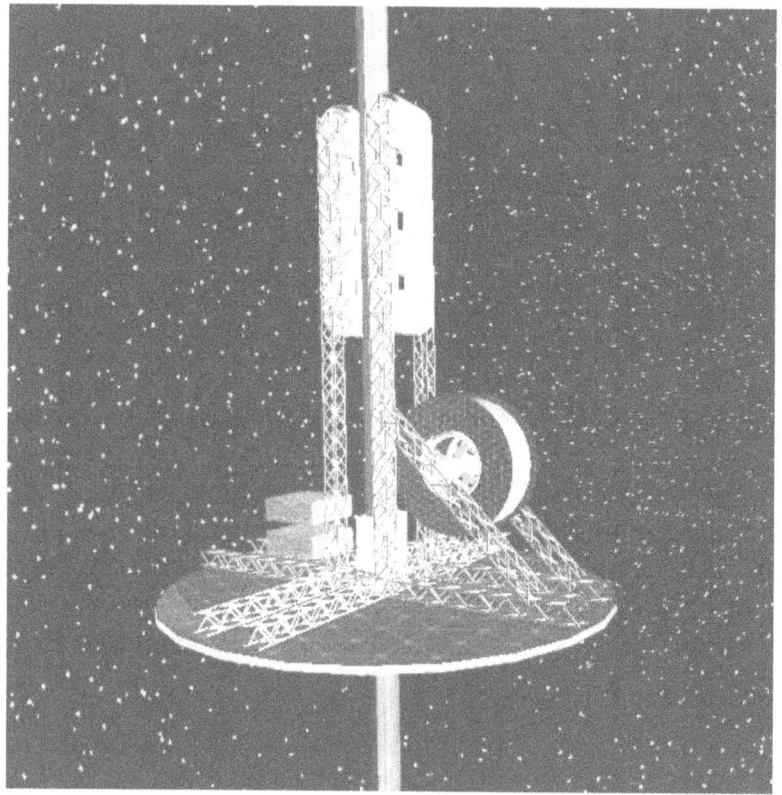

Figure 3-5. Laser receiver design [Edwards release]

3.4.3 Energy Management

As discussed earlier, ISEC expects that tether climbers will be powered by laser or solar energy. A laser receiver would be sizable while the solar energy alternative could provide 4 MW of power from an array with 100,000 m^2 of panel area weighing 1,220 kg, assume 40% conversion efficiency.

Energy storage must also be provided to sustain operations on the tether climber when the climber is shaded by the Earth. ISEC's baseline proposal is for 188 kWh of storage in a lithium-ion battery module with a mass of about 400 kg. This proposal assumes that ongoing research exploring the use of carbon nanotubes for battery anode and cathode structures will produce commercially viable batteries with an energy density of 470 Wh/kg. Such batteries are also expected to be used in electric cars and trains, to store energy from solar and wind power generators, and for various other terrestrial and space applications.

Power density—the amount of power that can safely be drawn from a battery of a given mass—is also a critical limiting factor. The best current lithium ion batteries have a power density of about 1,500 W/kg. Assuming the availability of carbon-nanotube technology, the government of

Japan has established a target power density of 2,000 W/kg for electric and hybrid cars by 2020, and 3,000 W/kg for hybrid trains by 2030 [TSM, 2010].

At that point, the 400 kg battery subsystem in the proposed tether climber could provide a peak power of 1.2 MW, enough to sustain a low-speed climb for brief intervals at low altitude, or full-speed climb through the brief interruptions in solar power expected at higher altitudes.

3.4.4 Communications Payloads and Antennas

As with other spacecraft, a tether climber needs to be in continuous communication with the space elevator operations center. ISEC expects that the climber will communicate with the GEO node, which will maintain contact with the Marine Node below and with company headquarters.

ISEC expects that Ka-band microwave will provide the primary high-rate communication channels for all of these links, with S-band microwave for emergency backup. The Ka-band subsystem would likely add significant mass to the tether climber, which we estimate at 107 kg. The S-band subsystem will have negligible mass—only about 6 kg, including antenna, diplexer, receiver, and transmitter.

An alternative, to save mass, would be to use laser communications for the primary link between the tether climber and the GEO node. This solution will work only above the atmosphere (up to about 40 km). The mass and power savings could be significant; the mass of a laser communications subsystem on the tether climber is estimated at only 15 kg.

3.4.5 Command and Data Handling (CADH)

This subsystem includes the primary tether climber controller, a computer that processes commands from the operations center and communicates with the drive subsystem, sensors, actuators, and other active elements of the tether climber. The tether climber will have a triply redundant CADH subsystem to avoid a single-point failure mode. The estimated mass of the subsystem is 30 kg.

3.4.6 Environmental Control

Thermal management is a major problem in space. The side of the tether climber that faces the sun may reach temperatures of 250° F (121° C), while the shaded side faces the cold of space, as low as -454° F (3° K or -270° C). The vehicle is surrounded by vacuum for most of its travel, which is similar to being trapped inside a thermos flask.

In this situation, like any other spacecraft, the tether climber would heat up unless excess heat is removed by radiating it into space. This is especially critical for the tether climber given the megawatts of heat generated when the tether climber is descending away from the GEO node (in either direction). Removing this heat will require substantial radiators facing the shaded side of the climber.

Traditionally, this is done by circulating a refrigerant between the vehicle and the radiators. The International Space Station's active thermal control system (ATCS), which uses this approach, was in the news in late 2013 when one of the pumps for its ammonia-based refrigerant failed and had to be replaced. For the proposed baseline tether climber, the thermal management subsystem is expected to have a mass of 120 kg.

3.4.7 Attitude Determination and Control

Control of the climber will be relatively simple compared to current spacecraft, as one axis is defined by the tether path. The primary control variable for the tether climber as a whole will be the rotation around this axis. Other control factors, such as balance trim, will be relatively minor.

In other spacecraft, attitude control is provided in various ways, including spin stabilization, reaction motors (typically cold-gas thrusters), and reaction wheels that are spun up or down to create an opposing torque on the rest of the vehicle. Spin stabilization is not an option for the tether climber, and the fuel requirements for reaction motors would likely make them impractical in this application.

A reaction wheel may be the only way to provide adequate control authority for a vehicle as large as the tether climber (up to 20 metric tons with cargo), but such a system would be fairly large and heavy. Another drawback is that because reaction wheels can only spin up to a certain speed, they can become unable to provide any more torque in a given direction, a condition known as saturation.

To solve this problem, ISEC favors the use of magnetic torque rods, a solution that is also used on the Hubble Space Telescope (where they are known as magnetic torque bars). Applying current through a wire coil creates a magnetic field that interacts with the Earth's magnetic field to produce torque. Although the peak torque is lower than that available from reaction wheels, it can be applied indefinitely.

ISEC believes that magnetic torque rods alone may be sufficient for rotational attitude control in the tether climber. If not, they may be used in conjunction with a single reaction wheel (vs. the three wheels that Hubble needs for three-axis stabilization). Torque rods would provide a way to bleed off the rotational momentum in the reaction wheel, preserving its ability to control the tether climber's rotational position.

Due to tether curvature, the space elevator will not be exactly vertical, with probably no more than 20 degrees of pitch angle at any point. If solar arrays are used, however, they must point at the sun for optimal results. Similarly, a laser power receiver array will need to be aimed at the laser power transmitter. As a result, these subsystems will need a solution for pitch control.

Various sensors would identify the location of the sun, identify the hot and cold parts of the spacecraft, and understand the radiation levels being encountered. With multiple GPS sensors, both location and orientation can be computed from differential measurements. Once the spacecraft is oriented correctly, the direction to the sun can be identified, allowing effective pointing of the arrays.

Solar panels would be hung below the tether climber with semi-rigid rods separating them vertically and small cables connecting them to change the up-down angle as the sun goes through its daily motion with respect to the tether axis. The array would point below the horizon at sunrise, point upward at local noon, and follow the sun down again until sunset.

Similarly, the orientation of a laser energy receiver would be maintained so that it points downward toward the transmitter, which would be located near the Marine Node. ISEC estimates that the mass of the tether climber attitude control system at 60 kg for computers, sensors, and the magnetic torque rods, plus 300 kg for a reaction wheel if needed.

The following table gives an estimate of the space elevator tether climber mass.

Subsystem	Mass (kg)	Notes
Payload	555	Robot arm, strap-downs, protection
Structure	1,500	20m in diameter x 15m in height
Thermal	120	Environment control
Power	2,340	Arrays (1,941 kg) + batteries (399 kg)
TT&C	180	30 CADH, 107 comms, 43 cable +
ADCS	360	60 sensors, 300 reaction wheel
Propulsion	740	Estimate using CNTs
Dry mass	5,795	
Propellant	205	Standard load for on-tether operations
Wet mass	6,000	

Table 3-6. Tether climber mass breakdown, assuming use of CNT materials. (Adapted from Table 3-5, with additional notes.)

3.5 Motor / engine and drive apparatus

The last section of the discussion on operational climbers is about the drive train of the climber to include the electric motor for the climber's wheels. Linear motor drives are most suitable for the tether climber. In 2003, a linear motor car achieved a speed of 581km/h during its manned test run in Yamanashi Prefecture.

Tsuchida et al. [Tsuchida, 2009] have discussed the optimization of vertical linear motors for tether climbers. On that method, the climber elevates by the rollers' friction force. Tether climber research and development has been carried out around the world investigating friction drive and optimization of friction drive mechanisms. This friction method is also considered an effective means for the initial space elevator. One example that was shown by Highlift in the early part of the last decade was a series of wheel segments as shown in Figure 3-6. [Laine, Chapter 3]

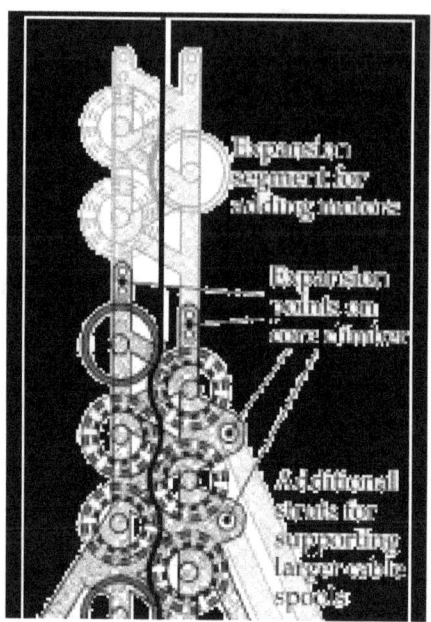

Figure 3-6. Climber Drive Apparatus [HighLift, Ch. 3]

This type of mechanism is simple and easy to manufacture. It can be applied to tether climbers. At Japan's Space Elevator Technical and Engineering Competition 2011 [JSETEC, 2011], the tether climber averaged about 17m/s (60km/hr) on a rope tether and about 9 m/s on a belt tether. This proved that roller friction drives could be successful. At the European Space Elevator Challenge 2011 [EUSPEC, 2011], they asked the competitors to calculate exact energy efficiency. The material is selected taking into account the various elements, coefficient of friction, wear resistance, use in outer space, aging and so on. The coefficient of friction is the most important; but it also depends upon the pressing mechanism, contact area, etc. Tsukiyama clarified the tribological properties of CNT film by experiment. He produced a CNT film. The interface strength was 85 GPa when the length of the CNT was 100 nm [Tsukiyama, 2010]. In addition, Umehara [Umehara, 2007] obtained knowledge of CNTs' coefficients of friction through testing. Much research needs to be accomplished within this field.

3.5.1 Wheel Interface Characteristics

There are many approaches for drive mechanisms that could work with a tether [Lades, 2013]. Some of the categories of tether-climber interfaces are shown in Figure 3-7, including:

- Contactless (fields instead of mechanical touch point – maglev example)
- Mechanical Contact – Wheels
    - Capstan (tether is wrapped around solid wheel increasing contact area)
    - Pinching wheel (flexible wheel is deformed to increase contact area)
    - Tracks
    - Hybrids

- Mechanical Contact – no wheels
    - Walker
    - Inchworm
    - Tether-vibration drive

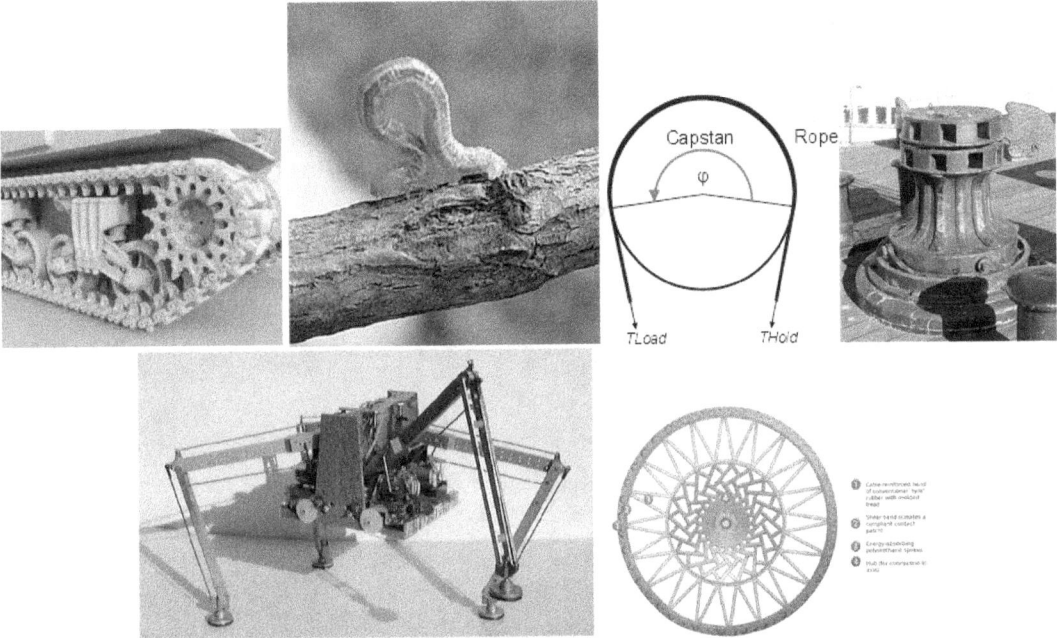

Figure 3-7. Types of mobility devices [Lades, 2013]

Dr. Lades conclusion during his presentation at the 2013 International Space Elevator Conference was that "Non-wheel climbers are less viable than wheeled climbers." However, during a brainstorming session at the same conference, an innovative concept surfaced and was discussed. This new approach is a combination of the inchworm and a walker. The concept is expanded upon in Appendix C.

3.5.2 Friction

Friction is a complex systemic phenomenon. There are several types that must be addressed during the discussion:

- Coulomb Approximation (dry atomically close contact on small fraction of surface, contact area is proportional to normal force until saturation & not exactly linear)
- Adhesive tape (no force, proportional to area, similar to drag racing tires)
- Gecko/Nanotubes, or active control of adhesion
- Inelastic process (internal friction results in head and mechanical wear/fatigue of wheels and the tether)
- Maximizing friction by increasing the active surface
    - Wheel size
    - Pinched wheel

- Capstan

During the discussion on wheel interaction with a tether, two topics must be understood. They are heating and fatigue. Heat is created when rotating wheels are deformed by the pressure of contact with the bearing surface. Figure 3-8 shows this deformation process in the context of a space elevator drive wheel bearing against the tether.

In space, convective cooling is not available. Thermal energy can only be removed from the wheel and the tether through conduction and radiation. The fatigue issue is related to inelastic wear per wheel rotation and occurs when there is an exchange/loss of material between the climber and the tether. Both the heating and fatigue issues "appear possible to be addressed with proper choice of wheel size (limiting the number of turns)." [Lades, 2013]

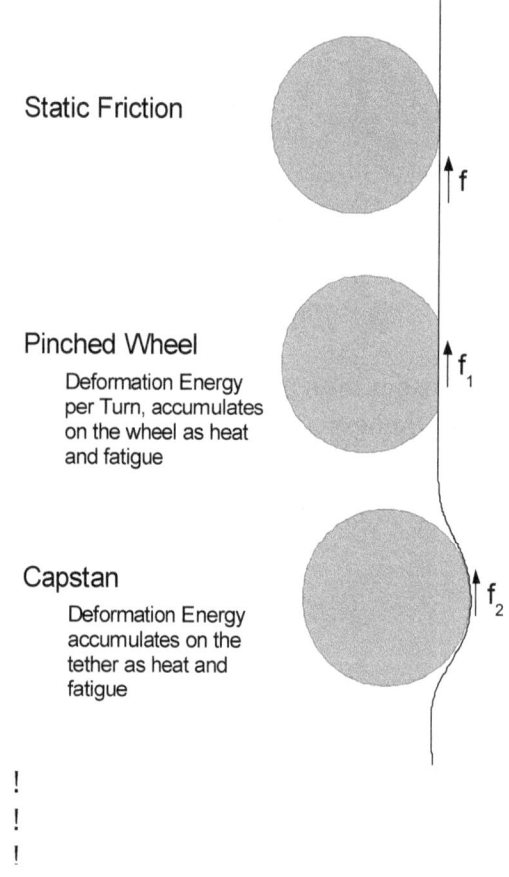

Figure 3-8. Pressure between wheels and their bearing surfaces result in the deformation of both objects, which produces heat as wheels rotate. [Lades, 2013]

3.6   Tether Construction Climber – Initial Design

The following discussion, adapted from [Bartoszek, 2013], illustrates some key aspects of the tether climber gripping and climbing activities. By having Mr. Bartoszek focus upon a simple climber (initial construction climber), many of the key attributes will be illustrated and initial designs shown.

The design of the construction climber starts with the most general requirements. The climber is supported by the ribbon using friction alone. The support from friction comes from squeezing the ribbon between powered rolling elements and the coefficient of friction between the rollers and the ribbon. The design of the traction drive in Edwards and Westling's book uses a tread wrapped around wheels similar to a tank's design. It is easy to show mathematically that the tread is not a useful addition to the design and must be eliminated.

Our original conceptual design used pairs of wheels clamped to each other with the ribbon in between. The parameter that determines how tightly the wheels must squeeze the ribbon is the coefficient of friction between the ribbon and wheels. If the ribbon is too slippery it may not be possible to squeeze it hard enough to develop the traction needed to lift the climber. The theory starts with a free body diagram of a single wheel pair pushed against the ribbon as shown in Figure 3-9. The wheel on the opposite side of the ribbon is not shown, but it provides the symmetric reaction force to the compression force F. (The wheels are clamped to each other around the ribbon through the structure of the climber.)

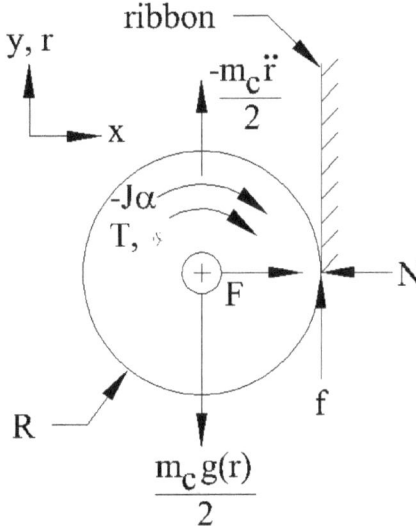

Figure 3-9. Free body diagram of a wheel pressed against the space elevator ribbon. The force F is balanced by the reaction force coming from a second wheel (not shown) on the opposite side of the ribbon. [Bartoszek, 2013]

The free body diagram neglects the losses from bearing friction and rolling friction. This model assumes that the entire weight of the construction climber is supported by a single wheel pair. If the climber is supported by more than one wheel pair, the weight is assumed to be distributed equally between all the wheel pairs, so the m_c term (climber mass) would be divided by the number of wheel pairs. This is similar in concept to the addition of locomotives to pull ever-longer trains. An equation of motion for the climber can be written by summing the moments around the point of contact between the wheel and ribbon.

$$\sum M = T - \frac{m_c \ddot{r} R}{2} - \frac{m_c g(r) R}{2} - J\alpha = 0 \qquad (1)$$

Where:

M = moments summed around point of contact between wheel and ribbon
R = radius of the wheel
N = normal force between ribbon and wheel
F = applied force compressing wheel to ribbon
T = applied torque from drive train
m_c = mass of the climber = 900 kg
f = friction force between ribbon and wheel
g(r) = gravitational drag force expressed as a function of r, radius from the center of the Earth

$$g(r) = \frac{M_e \cdot G}{r^2} - r \cdot \omega^2$$

G = Newton's gravitational constant

$$G = 6.67 \cdot 10^{-11} \cdot \frac{m^3}{sec^2 \cdot kg}$$

M_e = mass of the Earth

$$M_e = 5.9788 \cdot 10^{24} \cdot kg$$

ω = angular velocity of the Earth about its axis

$$\omega = 7.2929 \cdot \frac{10^{-5}}{sec}$$

J = rotary mass moment of inertia of wheel, kg-m²
α = rotational acceleration of wheel, sec⁻²
$\ddot{r}$ = linear acceleration along ribbon
x, y = Cartesian coordinates, y along ribbon, x perpendicular to face of ribbon
θ = Angle of rotation around the axis of the wheel, radians

The contact is assumed to be rolling and not sliding, so the linear and angular positions, velocities and accelerations are related by the following expressions:

$r = R\theta$,
$\dot{r} = R\dot{\theta}$ and
$\ddot{r} = R\ddot{\theta} = R\alpha$

Rearranging the terms of equation (1) and making the appropriate substitutions gives an equation for the torque required to accelerate the climber upwards with any given linear acceleration :

$$\ddot{r}$$

$$T = \ddot{r}\left(\frac{J}{R} + \frac{m_c R}{2}\right) + \frac{m_c g(r) R}{2}$$

(2)

The left side collection of terms (to the right of the equal sign) are the inertial terms. These give the torque required to accelerate the construction climber from any given initial velocity to any final velocity. The acceleration will be very small. Only 1% of the total torque near Earth is needed to accelerate the climber to 60 m/s (216 kph) in 10 minutes. The second expression in the sum is the torque required to keep the climber at constant speed. It can be thought of as the braking torque required to hold the climber on the ribbon at a point and not let it descend. This is one of the conceptual differences between an electric car on earth and the climber on the space elevator. Typically on Earth the car is not required to exert a constant torque just to stand still. This second term is critical with real motor characteristics because it only goes away at GEO. In constant velocity analyses, the r-double-dot acceleration term is zero so the right term is the only one to consider. As the climber rises, the pull of gravity declines with an inverse square relationship, but for thousands of kilometers near Earth, the climber must exert a significant braking torque.

3.6.1 The Mass Budget of the First Construction Climber and the Mass Problem of the 2004 Conceptual Design

A rendering of the 2004 conceptual design is shown in Figure 3-10 (from 2004 International Space Elevator Conference). The design of the ribbon construction climber assumes the mass distribution from Table 3.2 of "The Space Elevator" by Edwards and Westling [Edwards, 2003] as the design goal. One goal of this discussion is to determine how the original conceptual design of the drive system can be modified so this re-design is within the mass budget in the table of less than 233 kg. The number 233 kg comes from adding the masses of the motors, track and rollers, and structure in the table. (The drive system must be less than this number because the entire structure budget cannot be consumed by the traction drive system alone.)

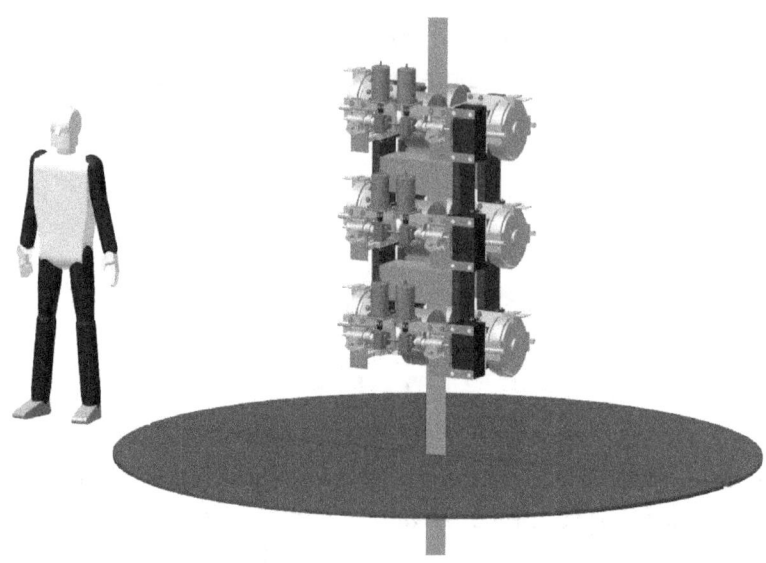

Figure 3-10. Overall view of the original conceptual design presented in 2004. The design is made up of three floating axle drive modules and three-fixed axle drive modules. The drive modules are connected to each other by structural weldments. The blue circular segments represent the photovoltaic (PV) planes. No structure is shown connecting the traction drive and the PV plane. [Bartoszek, 2013]

The bottom line of the conceptual design from 2004 is that it could not satisfy the mass budget. It was too heavy by a factor of 2.4 with 20 kW motors and a factor of 3 with 50 kW motors. As Table 3-7 shows, the motors represented almost 56% of the 233 kg budget for the drive train.

Component	Mass (kg)
Ribbon [buildup tether to disperse]	520
Attitude Control	18
Command	18
Structure	64
Thermal Control	36
Ribbon Splicing	27
Power Control	27
Photovoltaic Arrays (12 m^2, 100 kW)	21
Motors (100 kW)	127
Track and Rollers	42
TOTAL	900

Table 3-7. Mass distribution of components of the first construction climber [Bartoszek, 2013]

Table 3-8 shows that the mass of the motors found in 2004 made up only 13% of the total mass of the design, yet represented two thirds of the allowed budget. The fact that the motors used were lighter than the budget for the traction drive meant that the structure was the problem in reducing the mass of the drive. The mass of the conceptual design without the motors was 562.5 kg, but the budget for these components was less than 106 kg. The structure needed to be reduced in mass by a factor of 5.3.

Components in climber traction drive, assuming 20 kW motors	Component mass (kg)	% of total mass
Self-aligning bearings (12)	16.2	2.51%
Axles	32.1	4.96%
Interface structural material	51.2	7.91%
Wheels (6)	52.7	8.15%
Schmidt couplings (6)	62.6	9.68%
Structure for fixed wheel modules (3)	70.7	10.94%
Motors (6)	84.0	12.99%
Compression mechanisms (3 pairs)	136.1	21.04%
Structure for floating wheel modules (3)	141.0	21.81%
Total mass	646.5	100.00%

Table 3-8. Mass breakdown for the major components of the conceptual design [Bartoszek, 2013]

As the table shows, the largest contributions to the mass of the traction drive are in the last two lines. Almost 43% of the mass of the conceptual design is in the compression mechanism and structure of the floating wheel modules. Reducing the mass of this part of the traction drive has the biggest effect on the total mass. The structure will have to be reduced in cross-section and made of a lighter and stronger material to achieve the weight reduction. As the original structure of the construction climber was not analyzed for its structural efficiency, it is possible that significant reduction is possible while keeping the material aluminum. Such engineering solutions as "lightening holes" could be analyzed in the future.

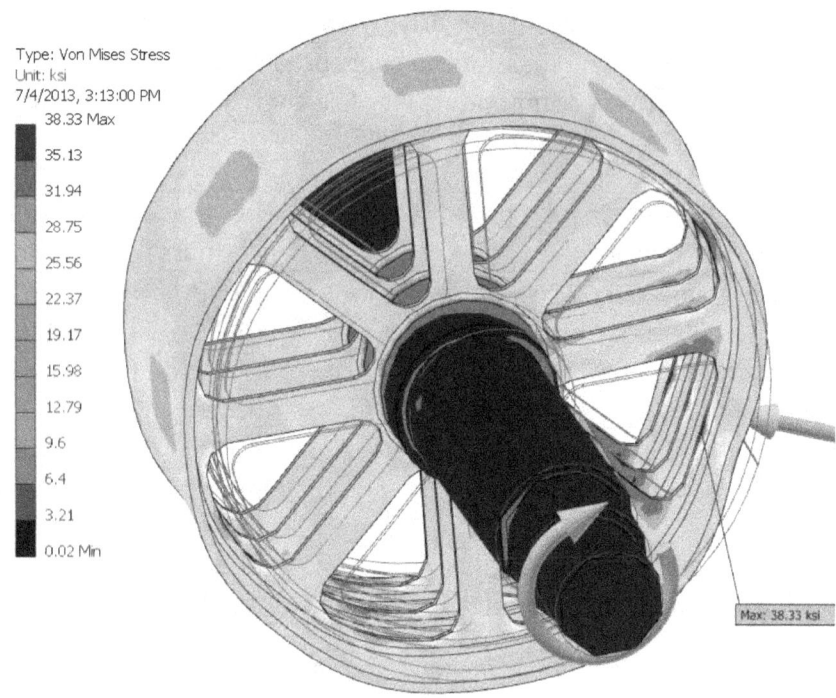

Figure 3-11. Plot of Von Mises stress for a wheel spinning at 10,000 RPM with the compressive load of 3,333 lbs. applied. The maximum dynamic stress is 38.33 ksi on the rim at the point of contact. [Bartoszek, 2013]

3.6.2 Conclusion on Construction Climber Design

It was not possible to definitively determine whether the 2004 conceptual design of the first climber can be lightened enough to satisfy the mass budget in Edwards and Westling's book. Further work has shown that there is a narrow range to the practical diameter of the wheels on the climber. Wheels that are too small would have to spin too fast and too many times to lift the climber at useful speed. Wheels that are too large cause the braking torque of the climber to be too high for practical motors, and their dynamic stress becomes too large for real materials. The construction climber will need more power than previously estimated to climb at useful speeds. It will require clever and detailed engineering to design motors in the appropriate power range that are light enough to satisfy the mass budget. Right now nothing certain can be said about the mass of a 100 kW motor for the Space Elevator.

Fatigue of the rolling elements of the construction climber must be in the mind of the designers at all times. The 100,000 km ribbon requires the climbers to live most of the life of an Earth electric car within one month and one trip with no repair stations. The structure around the compression drives of the floating axle modules in the 2004 design is where the bulk of the structural material is and where the greatest effort is needed in redesign to lower this mass. The drag force of gravity pulling the climber down the ribbon drops off very quickly with altitude up the ribbon. Given the very large power and torque capability necessary near the surface of the Earth, the construction climbers may be able to accelerate well past 200 km/hr at the higher altitudes, as long as the fatigue strength of the rolling components is not exceeded. In addition, the heating of the spinning masses will have to be assessed.

3.7 Further Thoughts about Tether Climbers

During the 2013 International Space Elevator Conference, active brainstorming occurred during the climber mini-workshop. The activity was designed to stimulate new and exciting ideas beyond the baseline discussed up to that point. Several panels jumped into the brainstorming approach aggressively and brought forth ideas and then reported them to the full conference. Another climbing and ribbon attachment approach was presented that answered an engineering need: to minimize the impact of the climber upon the tether. The wear and tear of multiple wheels gripping the tether at high speeds could be deleterious to tether life. The concept of hand-to-hand climbing should minimize the impact upon the tether. This alternate approach was refined during the conference and will be presented, in preliminary form, as Appendix C: An Alternate Climbing Approach.

In addition, several other new ideas surfaced that have great promise to better understand space elevator tether climber systems design approaches. Even though there are overlaps, initial results will be shown by panel. This maintains the group dynamics from the conference and is a better explanation of the concepts and their derivations.

3.7.1 Panel A

This panel looked at various concepts for what the climbers would accomplish. Some key points were summarized as: there is no such thing as extra power; all things brought up from the Earth have value; and most climbers will hand off payloads. During the panel discussions, eight different types of climbers were described:

- Construction
- Atmospheric (up and down – dock on stratoballoons?)
- First 7,500 kms (high gravity)
- 7,500 to GEO (1 MW power standard)
- Beyond GEO
- Personnel Climbers
- Cargo Climbers
- Micro to macro small climbers using the tether (transparent to other climbers)

3.7.2 Panel B

The initial climbers will be used to build up the tether mass. This will lead to a natural evolution of climbers with the following major points:

- Construction climbers need to incrementally build stronger tether capacity (20 to 40 to 60 MT, etc.)

- A second tether must be put in place prior to commercial operations from the first.
- Repair climbers are important, from LEO to GEO, for monitoring tether health, as well as monitoring debris fields
- Payload climbers should also be able to de-orbit
- Astronaut driven climbers must be able to be rescued (detachable, re-enterable)
- Low orbit options to include family trips and altitude freefall records
- GEO tourist climbers will be much larger with extra consumables
- A logical answer would be to feed energy from GEO solar power satellites to climbers, both above and below GEO
- Climbers dedicated to launches to Mars and beyond will include a Mars elevator and power projection to surface.
- In the far future, Mag-Lev options could work with a large Apex Anchor (asteroid)
- Ring World is reasonable for GEO stations connected.
- Outer planet moon trips reasonable in the future

3.7.3 Panel C

The movement from tether to mission orbit will be analyzed often as there are many variables. Some of the insights are:

- Circular orbits will probably need extra thrust once released (not in GEO)
- Elliptical orbits can result from almost any altitude release
- One question to pursue in the dynamics simulation is how much does the motion of the tether affect the release velocity (vector, direction and magnitude)
- One concern is to figure out how to coordinate up/down climbers on same tether
- On a single strand, to go around another climber, someone must disconnect/re-connect – maybe robotic arms?
- Below GEO, the process would be something like: stop, disconnect, activate thrusters to enable mission orbit when significantly away from tether.
- Most situations are very dynamic and release of climber will affect mass on tether
- To lower the chaos in launch from the variable dynamics of the tether, many perturbations could be added to counter movement for debris avoidance, releasing and attaching cargo, varying climber speeds and any unpredictable movements.

3.7.4 Panel D

This panel approached tether climbers with the idea that the various altitude regimes would demand optimization in designs. This led to a list of reference points to climbers specialized for different altitudes:

- There are strong reasons for varying designs of climbers.
- Low-level climber will probably be driven by laser.
- Higher speed will come from solar at higher altitudes.

- One approach for multiple designs is to have hand-offs. Each climber goes up, hands off payload, goes back to lower range, and picks up the next payload.
- The handoff approach could impact the total throughput by about half.
- Only four climbers would be required on single tether at any one time… payloads are passed along.
- To repair change-out climbers, three high ones come down to lower climber that then loads them as payload and descends to surface.
- After proof test of concept, larger advanced climbers can replace original concept and could go from LEO to GEO.
- This panel of the mini-workshop concluded that four separate optimized climber designs would be required, but this must be confirmed by a more detailed design and operational study.

3.8 Tether Climber Conclusions

The space elevator tether climber can be built with today's technology. However, there are many questions and challenges that will ensure excellent designs will have to be developed leveraging historic spacecraft technologies, new technologies to emphasize strength and lightness, as well as understandings of the tether material properties that are not available today. One strength is that there are 60+ years of heritage in spacecraft design that can be utilized when structuring tether climbers.

References

- Bartoszek, Larry, "Getting the Mass of the First Construction Climber Under 900 kg, International Space Elevator Conference Seattle, 2013.
- Edwards, B. C., and Westling, E. A., "The Space Elevator: A Revolutionary Earth-to-Space Transportation System", BC Edwards, 2003.
- EUSPEC (2011),"Evaluation" at: http://euspec.warr.de/handbook
- JSETEC (2011), "Results" at: http://www.jsea.jp/ja/jsetec2011_result
- Lades, Martin. "Climber-Tether Interface for a Space Elevator," International Space Elevator Conference Seattle, 2013.
- Laine, Michael, Chapter 3: Spacecraft at: http://www.mill-creek-systems.com/HighLift/chapter3.html
- Shelef, B. (2004), "Segment Based Ribbon Architecture"., In Proc. of 3rd International Space Elevator Conference, June 2004.
- Shelef, B. (2011), "The Space Elevator Feasibility Condition", Climb Journal, Volume 1, Number 1, p. 87.
- Shelef, B. (2008a), "Space Elevator Power System Analysis and Optimization, Spaceward Foundation, 2008. Available at: http://www.spaceward.org/elevator-library#SW
- Shelef, B. (2008b), "The Space Elevator Feasibility Condition", Spaceward Foundation, 2008. Available at: http://www.spaceward.org/elevator-library#SW

- Shelef, B. (2008c), "A Solar-Based Space Elevator Architecture," Spaceward Foundation, 2008. Available at: http://www.spaceward.org/elevator-library#SW
- Suemori, K. (2012), "Film-shaped thermoelectric conversion elements can be produced in print" Available at: http://www.aist.go.jp/aist_j/aistinfo/aist_today/vol12_04/p17.html
- Swan, P., Raitt, Swan, Penny, Knapman. International Academy of Astronautics Study Report, Space Elevators: An Assessment of the Technological Feasibility and the Way Forward, Virginia Edition Publishing Company, 2013.
- TSM (2010), "Technological Strategy Zmap 2010 – Energy", Ministry of Economy, Trade and Industry. Available at: http://www.meti.go.jp/policy/economy/gijutsu_kakushin/kenkyu_kaihatu/str2010download.html
- Tsuchida A. et al (2009), "New Space Transportation System-Space Train (Elevator) : World trends and Japanese Space Train Concept", Technical report of IEICE. SANE 109(101), 93-98, 2009-06-18
- Tsukiyama, Y. (2010). "Tribological properties of high-alignment carbon nanotube films", The Machine Design and Tribology Division meeting in JSME 2010 (10), 49-50, 2010-04-18
- Umehara, N. (2007), "Tribology of Fullerane and Carbon Nano Tube as Advanced Materials Designed Nano-structures", Shinku, Vol. 50 No. 2, 2007. Pp. 76-81
- USAF (2012), "Energy Horizons", United States Air Force, Energy S&T Vision 2011-2026, AF/ST TR 11-01 31 January 2012, Pgs. 21-24.

Chapter 4   Power Sources

The customer needs center around routine and inexpensive delivery from the Marine Node to GEO within a week on a smooth and reliable climb. With this thought in mind, calculations have shown that during the heavy gravity region, the minimum need is between 2.5 and 4.0 MW of power to the climber for storage into the battery and for "pulling" the climber up the tether [Knapman 2013]. This chapter compares and contrasts the two accepted approaches of laser power from the surface of the ocean and solar power during daylight hours. The last section of this chapter combines the strengths of both into a "hybrid climber" concept with lasers during the high gravity portion and solar for the rest of the climb. This leverages the best of both concepts and ties together a tether climber that just might be the right answer. There have been many studies on the topic (references are provided at the end of chapter): this chapter will explain the two choices and compare them. There is a third option of microwave energy that has future potential but for this study is not considered. Table 4-1 shows the options for power in the different operational regions.

Operational Region	Power Option	Power Required
Atmospheric	Power cable or laser	4 MW
40 km-GEO	Laser or solar	4 MW
Beyond GEO	Solar	Minimum
Return from Apex Anchor	Solar	Moderate
Return from GEO	Solar	Minimum
Repair	Laser or solar	Less than 4MW for smaller
Construction	Laser or solar	Lighter so less power

Table 4-1. Power Requirements on Climber 4 MW constant power – per Region

When talking about the power required for a space elevator climber, the assumption is that the tether climber cannot carry stored energy, for climbing, in the form of fuel or nuclear sources. The mass of these power sources, with all the associated apparatus, would be too great to allow any reasonable payload to come along for the ride. Therefore, the basic requirement is that energy be supplied from outside of the tether climber in the form of radiated energy (solar or laser).

One of the major thoughts that will carry through this chapter is that the current technologies would not work effectively (probably would work, barely) to move the climber at a desired rate. However, there is tremendous research and development being conducted into power sources, power projection (laser energy to the climber), efficiency of receiving cells (at both solar and laser wavelengths), lightweight materials, and thermal pathways (see wiki references at end of chapter for solar arrays and for space elevators). This ISEC study will project some of the technologies to show that the future should enable the tether climber to receive sufficient energy and be able to climb with sufficient speed to have a weekly delivery of customer cargo at GEO. Two more thoughts are elemental: (1) obtaining high levels of energy on the surface of the Earth is complex and operationally taxing, and (2) solar energy is ubiquitous in space (except during eclipses). The actual selection of the best method to deliver power to tether climbers will be accomplished during the next 15 years as the space elevator matures in concept and the development begins.

4.1 Laser Power Option[1]

The baseline for Dr. Edwards' work focuses on several large lasers that project power to the tether climber from the surface of the Earth. His initial baseline is three free-electron lasers for each climber. To enable the concept being described in this study, Dr. Edwards' ideas must be expanded to include large lasers on the surface of the ocean separated by tens of kilometers. Each climber will require one or more laser, and additional lasers and power supplies will be required for backup. The massive energy required (per laser and times 7 tether climbers) could be supplied through multiple sources, to include nuclear power or generators. The beauty of the concept is that the total power to support tether climbers is controlled within the space elevator infrastructure and the movement of the climbers will be dependent on the terrestrial infrastructure. Figure 4-1 shows the concept of a laser power system working with the tether climber.

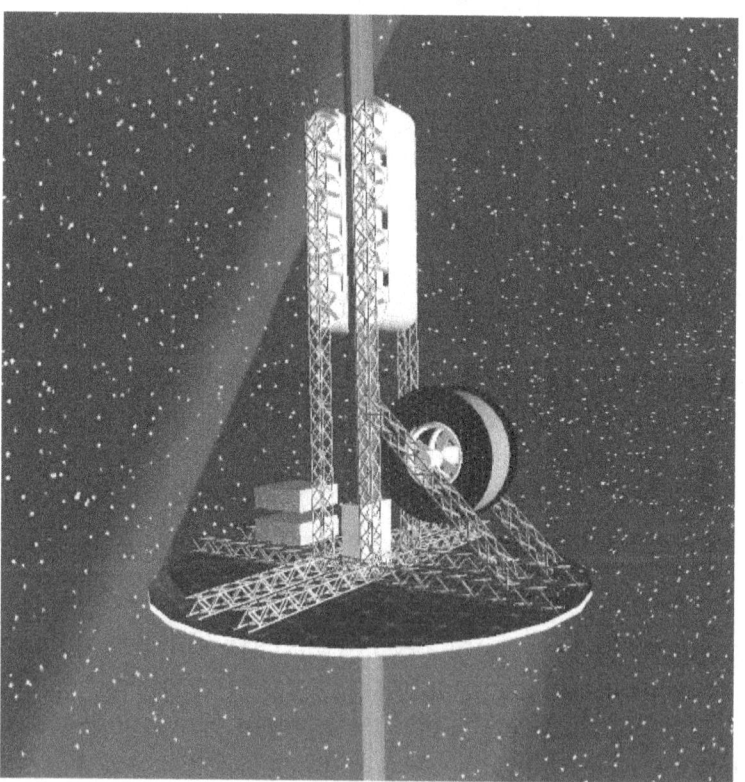

Figure 4-1. Tether Climber with Laser Power Source [Edwards & ISR, 2003]

---

[1] Much of this discussion, in section 4.1, is based upon discussions on the wiki, derived from Dr. Edwards's early work. http://spaceelevatorwiki.com/wiki/index.php/SpaceElevatorPower

Adaptive optics, a technology currently used in astronomy, can help mitigate atmospheric influence. Interference from clouds can be mitigated with more redundant laser sources. Each Marine Node can be paired with a single laser transmitter to start the climb. A second laser power transmitter would take over when the climber reaches a predetermined altitude to provide energy for the long run up the elevator. The latter transmitter could be positioned at the top of mountains within 100 km of the space elevator infrastructure. This would greatly increase the efficiency of the energy transmitted, although the best locations for the Marine Node are far from mountains.

The baseline of Dr. Edwards' study shows the feasibility of high power lasers at the Marine Node supplying energy to the tether climber. Some conclusions are (1) that the laser must use adaptive optics, (2) the energy required is quite high, and (3) the total efficiency is quite low when accounting for losses, even with diode lasers that are about 50% efficient. ISEC expects that there is an excellent chance that the technologies will mature into ones that can be leveraged for the space elevator infrastructure.

Table 4-2 shows the basic characteristics of a space elevator laser power transmitter.

	Value	Details
Power input per laser (MW)	20 - 150	Can be from any conventional source. Ocean going platforms can have this amount of power available from diesel engines.
Power in laser beam (MW)	2 - 15	This power does not need to be in a single beam but can come from multiple beams interlaced in time (pulsed beams)
Primary Optics		Segmented adaptive optics mirrors similar to those in major astronomical observatories
Primary Beam diameter (m)	10-30	This size is determined by focusing required intensity to a relatively safe level.
Wavelength	TBD	To be matched to the type of cells and lasers available.
Tracking accuracy	TBD	Determined by laser beam and receiver sizes.
Power source	Solid-state laser	Stacked diodes are also coming on line. See Ultrahigh-Average Power Solid-State Laser.pdf, MCDL LASE 2004 1f.pdf, LASE 2004 SSDL 2k final.pdf

Table 4-2. Laser power transmitter operational parameters [adapted from http://spaceelevatorwiki.com/wiki/index.php/SpaceElevatorPower]

4.1.1 Power receiver alignment

The main challenge in designing a laser-powered climber is focusing the energy on the laser receiver. Each panel's shape and position must be designed to point at the laser source, as the acceptance of best efficiency of the energy requires close to 90-degree incidence. The tether is not a straight line, but one with significant dynamical motion off the straight line. [Swan, 2013]. The magnitude of the resulting pointing error will require further investigation. Efficiency for pointing errors follows the cosine error rule ($\cos x \approx 1-(x^2/2)$) for small angles; for example, a 5-degree error still permits 99.6% efficiency.

4.1.2 *Power receiver efficiency*

One large problem for laser power is achieving the very high conversion efficiency of the receiving photovoltaic array. Dr. Edwards' specification was greater than 90% conversion efficiency to keep the temperature of the climber low enough that it will not require cooling. This level of efficiency does not exist yet. Radiators could be used if the conversion efficiency is lower, impacting the weight and attitude control issues.

4.1.3 *Engineering Reality Check*

Power must be delivered in a tight beam over tens of thousands of kilometers. Various methods were examined and laser power beaming was selected as an alternative for delivery of power from the Earth's surface. This is a mature yet still rapidly evolving technology. The next table looks at the various aspects of laser power sources and discusses the near term activities needed.

Item	Notes
Categorize the available laser systems	The baseline power delivery system uses a set of high-power lasers to send power up to a climber ascending the ribbon. Lasers have evolved dramatically over the last 10 years from few large systems, primarily chemical, to now a growing availability of free-electron laser (FEL) and solid-state systems. With the current rate of change of the solid-state systems, a high-power system of the size required will be ready before we are ready to build the elevator. We need someone to evaluate the different lasers, keeping in mind the capital and operational cost, overall efficiency of the laser, efficiency of the matching receivers, operational lifetime, laser size and options for combining beams (in time for pulsed systems, what losses with CW systems), weather considerations, and other factors.
Detail optics system	Although it has been mentioned, it is probably unlikely that we can point a laser directly up and deliver the laser power to a 10m diameter spot 10,000 km away - we will likely need optics, no doubt adaptive optics. What we need someone to do is research the possible large optical system (astronomical telescopes) and get a full rundown on what they will do and what are the challenges for us. We need to know what it takes to get a large laser beam into the optics, what stability is needed, whether there are issues with operating this in a sea-spray environment, cost, complexity, etc.
Weather concerns for the power beaming	Initial studies of the selected anchor location show minimal clouds, minimal lightning, and minimal wind. What other issues are there for the power systems being considered? What is the down time that we can expect and realistically see? We need someone to start pulling this together. We also need evaluations of other proposed sites such as those proposed by Phil Ragan off Australia and the Japanese in the Western Pacific or off the Maldives.
Safety issues with high power laser	High power lasers can cause damage and injure people. We need to be able to operate this system safely and the question is how. Does everyone on the station need to be indoors behind closed windows? Or is there a scenario where a live person might actually be able to look at an ascending climber?

Table 4-3. Laser Power Status and Questions
[http://spaceelevatorwiki.com/wiki/index.php/SpaceElevatorPower]

4.1.4 Status

Boeing has demonstrated operation of a 25kW laser of the kind we need and claim they can build MW lasers. Beaming 9+ kW lasers over short distances (one km up) has been demonstrated at the Spaceward Space Elevator Games in 2009, as Figure 4-2 shows. Adaptive optics have been demonstrated with long distance capabilities.

Figure 4-2. Power Beaming in Action [http://www.spaceward.org/games]

4.1.5 Major Design Challenges

The primary difficulty in beaming laser power up to the climbers specifically from sea level is atmospheric distortion and absorption. Atmospheric distortion will broaden the beam and reduce the power delivery efficiency. From Electro-optical Handbook [Smith, 1993], we find a discussion of exactly the problem we are investigating—the problem of sending a laser beam from Earth to space. To beam up power from sea level we will require adaptive optics. From the work of Robert Fugate and others we find that adaptive optics (AO) have experimentally demonstrated a spatial resolution of 25 cm at 1000 km [Angel, 2000]. This is an order of magnitude better than our application requires at 1000 km and this system can focus the laser into the precise spot size we need at 10,000 km. With this accuracy we can place the power we need onto the 3-meter diameter laser receiver array we have designed into the smallest climber. By the time the beam expands to fill the photovoltaic array of our smallest climber (12,000 km altitude) the power requirements of the climber are lower due to reduced downward acceleration (~0.1g). The only thing that has not been demonstrated is the complete and continuous beaming of a high-power laser. Prior research has been based upon laser power transmission to space-based assets with little or no feedback capability.

The Space Elevator climber, as a cooperative target, will provide targeting data back to the transmitter thereby establishing a closed-loop system that continuously maximizes power on target. The primary problems that may be encountered in this next stage include thermal blooming of the atmosphere, and production of the high-power laser. In our application, thermal blooming will not be a problem with a large beam size and the power we will be using. The receiver system must also be considered when examining the power beaming system. There are several photovoltaic cells that can be used as receivers and the choice will depend on which laser is being used, cell mass and the desired operational lifetime. One additional problem that we need to address is lost transmitting time because of overcast skies. At our proposed anchor location where it would be best to also place the power beaming facility, the percentage of overcast skies appears to be low [Swan, 2013] but to ensure continuous operations a second beaming facility located in a separate weather zone would be advisable. In our proposed situation the second beaming facility could also be located on a movable ocean platform roughly hundreds to thousands of kilometers from the anchor or in the mountains of Ecuador (10,000 ft. altitude). An additional power beaming system in the United States' Mojave Desert [Bennett, 2000] could also be used for supplying power to climbers above 10,000 km.

One large concern by the operators of satellites is the risk of damaging down-looking optics. The Earth resources satellites all continuously collect data while staring at the surface with large optics. As evident by the recent activities around laser beaming towards pilots in flight, the concern is similar to operational satellites. As such, there is a process in place to ensure any lasers beamed towards space must gain approval prior to operations. The Laser Clearing House will ensure the beam does not impinge upon a satellite in operation, thus saving missions. The limited, current, activities with lasers looking up makes this process feasible today. However, if the community were to have seven large lasers beaming continuously towards the GEO arc, the concern would become real that dangerous situations could occur. A study to understand coordination of operational space flight and laser support from the ground must be initiated to understand the risks and the mitigation techniques.

In examining possible systems we see that there are performance, maturity and operational concerns. An overall summary of a laser power system is shown in Table 4-4. It is clear that at this time the laser power beaming system is a viable option, especially where it is most needed, deep in the gravity well of Earth. The higher efficiency, smaller transmitter, and maturity of the laser beaming system are all distinct advantages for a future concept.

Operating Wavelength	0.84 microns
Transmitter System	Free-electron laser / deformable mirror
Transmitter Area	12 m diameter
Receiver System	Tuned solar cells
Overall system efficiency	Low
High altitude operation	Preferred
Development level	Under construction
Number of lasers needed	At least seven for each tether
Power for each laser	Greater than 50 MW

Table 4-4. Laser Power System description [Edwards, 2003]

4.2 Solar Power Option

During the last ten years, the assumption was that the only power available would come from the surface of the Earth, as it was inexpensive and technologically feasible. However, during the last ten years of discussions, conference papers, IAA Cosmic Studies, and interest around the globe, many discussions have led some individuals to the following conclusions:

- Solar Array technology is improving rapidly and will enable sufficient energy for climbing
- Tremendous advances are occurring in lightweight deployable structures

As a result of these conclusions, a space elevator design team has shown that the advancements in solar arrays' reduced mass and power efficiency may enable this source of energy to provide climbing power.

As discussed in previous chapters, the solar power option especially matches the needs above the atmosphere and on cargo climbers where the time requirements are relaxed. The aerospace industry is actively pursuing extremely lightweight solar arrays with remarkable efficiencies for future spacecraft. The projections in this study utilize the advances projected for carbon nanotubes in both the lightweight but strong material and in their electrical characteristics. The future with carbon nanotube innovations is impressive. These research and development advances should enable sufficient power during the daylight hours to climb to GEO and beyond. During an eclipse, the tether climber will go into hibernation and be prepared to climb at dawn. This is only significant during the first night of climb as the angle to the sun improves rapidly as the shadow of the Earth recedes.

To enhance this design, the team has pulled together a systems approach to provide energy to tether climbers with ubiquitous solar energy:

- **Step One**: Launch at dawn above the atmosphere, currently estimated to be 40km high. As discussed in previous chapters, this report focuses on a constant-power implementation.
- **Step Two**: Use fragile solar arrays for tether climber power with large arrays strung out below the climber with pointing capability for sun searching. Figure 4-3 shows such a configuration.
- **Step Three**: Hibernate during eclipse. The maximum duration of hibernation is around 7 hours for first night and rapidly goes down from there to continuous solar impingement soon after the second or third day depending on solar season.

Figure 4-3. Solar Power Climber – Above atmosphere, prior to solar cell deployment

The enabling strength of the solar power option is that there is a capability to start the climb at altitude where the sun is supplying energy. The estimate is that if the climber were placed at 40km altitude, the significant effects of the atmosphere would be negated and the climber could deploy its large and fragile solar arrays, enabling the start of the climb. There are many ways to attain 40kms, as was described in the IAA study "Feasibility of the Space Elevator and a Way Forward." [Swan, 2013] Appendix A explains these approaches in more detail; however the main ones would be:

- *High Stage One*: A permanent base station at 40km altitude capable of supporting 400 metric tons without impacting the space elevator tether.
- *Extension Cord to Altitude*: This alternative is one that protects the tether climber inside a CNT box during the ascent in the atmosphere and then releases the deployed tether climber at daybreak. The extension cord runs the climbing motor and drive train and is extremely light with CNT materials for strength and even conduction of power.
- *Spring Forward*: This alternative takes advantage of the elastic properties of the space elevator tether. The tether is pulled down 40km, the tether climber is attached in a protective box, and the tether pulls the package up to 40km from the strain induced. The package is released for deployment and initiation of climb at daylight and the protective box is pulled down again for the next climber.
- *Laser Power to Altitude*: This is an alternative that would have a lower power laser working in the range of zero to forty kilometers on a one-on-one relationship for nighttime movement from the surface of the ocean to the starting point. The tether climber would be

enclosed in a protective shield for the environment and deployed at altitude as Figure 4-4 shows.

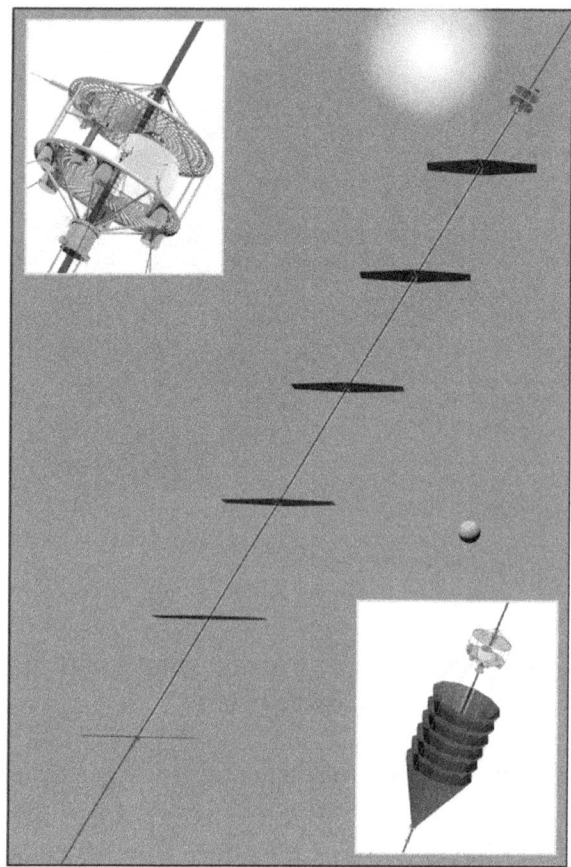

Figure 4-4. The tether climber (upper left) is shown with its solar arrays stowed below 40km altitude (lower right corner) and then deployed for ascending to GEO and beyond (center). [Chasedesignstudios.com]

The space elevator solar power architecture alternative has a basic 20 MT tether climber (6 ton structure, motor, power collector and a 14 ton payload capability) starting on the surface of the ocean in a one-gravity field. As it rises, the mass stays the same (because the energy source is external – not carried by the rider); but the force of gravity falls off as per the $1/r^2$ rule. The initial power requirement is estimated to be 11.8MWatts (electric) if the desire is to start off at high speed. This falls off by three fourths at a 6378km altitude or a 12,756km radius. Therefore, the critical power projection is near Earth with solar power sufficient along the path from start to finish.

Four factors dominate the discussion of motive power:

- This power alternative requires a constant power of at least 2.5 MW for slow climb at initiation with increasing velocity as gravity is reduced.

- Solar arrays should be able to accomplish the mission [Shelef, 2008c]. The projections of increases in solar efficiencies and decreases in weight are remarkable and may match future space elevator needs.
- NASA's solar array research has hardware in testing that is called "Solarosa" and provides: 400 to 500W/kg BOL (beginning of life) with 60 to 80kw/m^3. This would yield a mass requirement of 23.5 metric tons. Two notes for thought: BOL does not play, as the climbers are returned and can be refurbished, and this is 2012 technology with tremendous research going on for lightweight flexible solar arrays. If the technology improves by one order of magnitude in the next 15 years, the climber design for power falls into place. [NASA Tech Briefs, Nov 2012]
- The sun provides free energy without the complexity of generating power on the Earth or in orbit; space technology has been leveraging this "free energy" for the last 50 years. In addition, there is no pointing of lasers or adaptive optics required for atmospheric corrections. However, there is the standard need to point at the sun.

The proposed architecture for space elevator electric powered climbers is based on the following set of assumptions:

- Tether climber delivered to 40km altitude (Appendix A summarizes three methods for initiation of climb)
- Initial solar array power to provide energy from daybreak to first sunset (somewhere above 1370 km altitude)
- Tether Climber to "sleep" during eclipse
- Solar energy usage begins again at second sunrise and continues towards the space elevator Apex Anchor
- The bottom line is that this deletes the complexity of creating and transporting energy as a mission inside the space elevator infrastructure

While a set of several solar panels can give the climber large amounts of power, they can only do so when illuminated. Regrettably, at the equator, illumination only occurs 12 hours a day. It is advantageous to launch the climber, from its 40 km altitude starting point, at sunrise (05:35) so large amounts of power are available. When night falls, the best alternative is to wait until local dawn to proceed. Normally, the climber would have traveled far enough by nightfall that power requirements would have dropped significantly and the length of the night is reduced. An engineering design must be conducted to determine how heavy the power system would be and how far the climber has climbed after 24 hours. In both summer and winter, sunrise and sunset are coincident with the ecliptic plane, whereas in spring and fall, midnight and noon are. Sunset for the climber occurs after 20:30 on the first day, for two reasons: first, having traveled a certain distance, the climber is located at a higher radius so that it does not have to rotate 180° to clear the night side. Second, as a function of the sun's angle above the equator, the climber at sunset is located further from the Earth and, therefore, spends less time in the Earth's shadow. Table 4-5 shows some examples of climbing locations of the space elevator while in the spring/fall (worst case) scenarios. Two examples are shown: maximum 4 MW at up to 300 kph, and maximum 3 MW at up to 400 kph. The calculations have been run to show times and altitudes, assuming a start from 40 km altitude. All seven days are shown with sunrise and sunset and the arrival time at GEO that yields the full time for climbing. The dark eclipse time is calculated to show night duration, and sunset time is illustrated. The days are natural progressions, starting at 05:35 hours until dark. Day seven is shown, as that is when the tether climber reaches GEO altitude. Note that in this worst-case scenario, even at GEO, the climber will experience only a brief period of darkness – about 70 minutes.

Max power 4 MW, max speed 300 kph					Max power 3 MW, max speed 400 kph				
Altitude at night (km)	Hours of night	Eclipse angle	Sunset time	Sunrise time	Altitude at night (km)	Hours of night	Eclipse angle	Sunset time	Sunrise time
40	11.1	83.6	NA	5.57	40	11.1	83.6	NA	5.57
1370	7.4	55.4	20.33	3.72	965	8.0	60.3	20.02	4.06
3980	5.0	38.1	21.46	2.54	2465	6.2	46.2	20.92	3.08
9380	3.2	23.9	22.42	1.61	5060	4.5	33.9	21.75	2.27
15770	2.2	16.7	22.89	1.12	10590	2.9	22.1	22.54	1.48
22390	1.7	12.8	23.16	0.87	19225	1.9	14.4	23.05	0.97
29140	1.4	10.4	23.32	0.70	28165	1.4	10.65	23.30	0.72
Arrival time at GEO 23.04					Arrival time at GEO: 19.90				

Table 4-5. Duration of night at various altitudes for two power/speed scenarios. (All times shown in decimal hours.)

For the rest of the year, eclipse effects are less and less during climbing, until the winter and summer solstices, when the distance to full sunlight is at only 9,630 km. This can be reached in less than three days for a 4 MW climber. Between those two extremes of seasons, the path to full daylight varies. The conclusion is that the first night period will last about seven hours. The climber will emerge from the first night at approximately 03:30 and on the second night at approximately 02:00, after which the time of darkness will be little or nothing. A new climber starts each day at 05:35 from an altitude of 40 km. To reach 40 km, it may use box protection with a power cord from the surface, in which case the tether has to bear the load of the climber and box an hour or more before 05:35. If it uses High Stage One, the tether only has to bear the load of the climber from 05:35 hrs.

4.2.1 Solar power array suspension

The main challenge in designing a solar powered climber is the large and flimsy nature of the panels [Shelef, 2008c]. Typical in-space structures benefit from the lack of an atmosphere and gravity. In our case, we do not have the latter advantage and must design the structure to withstand 0–1g while being tiltable. As was shown in Figure 4-3, the basic concept has each panel's shape and position being completely determined by a large number of marionette-style pull-strings that suspend it against gravity. Except for the top panel, the panels are simply supported by the strings. Extremely lightweight (CNT) rods with extreme stiffness would be utilized. The distance between the panels is 3–4 times their diameter, corresponding to a shading angle of 14–18 degrees. This distance can be reduced at the expense of not getting optimal sun tracking around local noon. As the system has to function in zero g as well, a trailing "caboose" car keeps minimal tension in the strings when necessary.

4.2.2 Engineering Reality Check

A possible candidate for such a power architecture is the solar powered climber – a design capable of achieving sufficient and significant power levels required for climbing the tether [Shelef, 2008c]. The enabler for this design is the recent development of very low weight, thin-film, photovoltaic technology able to provide as much as 5kWatt/kg. Recently a German experiment developed a self-deploying solar panel demo weighing 32kg and able to provide 50kWatt of electric power under full sunlight illumination. The array size is 20m x 20m, and operates at slightly under 10%. The power density of the complete panel (foil and booms) is 1.6kWatt/kg. The next few years will see leaps in efficiency and reduction of the mass. The next table (Table 4-6) was taken from a US Air Force document [USAF, 2012] showing the expected increases in efficiency of PV arrays for spacecraft. The article states, "The importance of these S&T efforts lies in the fact that every 1% increase in solar cell energy generation efficiency translates to a 3.5% increase in power (or decrease in mass) for the system."

	Energy Generation	
Near (FY 2011-15)	Mid (FY 2016-20)	Far (FY 2021-25)
30-35 % efficient PV cells	40 % evolved PV cells	70 % efficient PV cells (e.g. quantum dots)

Table 4-6. Solar panel efficiency predictions [from USAF, 2012]

If this projection of future solar arrays is accurate, the improvement from 10% efficient panels (like the one described previously as a current prototype) to 40% efficient will provide phenomenal power. The increase in power would be approximately 105% (30 x 3.5% increase) or 104 kW for the same mass and area. This could also lead to an output power of approximately 50 kW with less mass and area. If the industry approaches 70% efficiency, the impact on the PV arrays would be a 210% improvement (approximately 60 x 3.5 with "quantum dots"). Table 4-7 (a copy of Table 3-4) shows the relationship between the current array prototype discussed and the needed size with the current array and two future cases.

	Current array	Scaled up for 20-ton tether climber	Future array, medium efficiency	Future array, high efficiency
PV efficiency	10 %	10 %	40 %	70 %
Output (kW)	50	4,000	4,000	4,000
Mass required (kg)	32	2,500	1,220	810
Area required (m²)	400	200,000	100,000	66,000
Size of square array	20 m x 20 m	447 m x 447 m	316 m x 316 m	257 m x 257 m

Table 4-7. Projected improvement in capacity. Note that the projections for mass are probably high, as the sparseness of the array will increase power and decrease mass required for the same level of output.

In the best-case situation, the arrays could be ten in a series with 26 m x 260 m panels looking at the sun in a very benign environment above 40 km altitude. One interesting note is that the development of ultraviolet solar cells would increase the efficiency significantly. The convenient aspect of this development is that there is a tremendous effort around the globe to increase the efficiency and manufacturability of solar cells and their associated equipment for both terrestrial and space uses. In addition, there is a tremendous effort to develop extremely lightweight materials for space applications. It is clear at this time that the solar power system is a very viable option, especially when it starts at 40 km altitude.

4.2.3 Laser Clearinghouse Requirements

- MUST have permission to radiate 24/7/365 – very difficult to obtain at extreme high power
- Risk damage to ALL satellites with their downward sensors such as: Earth Resources main telescopes, attitude sensors (Earth looking), and science sensors. The energy impingement on satellite components has not been designed for predicted levels. (It is high energy to climber at 30,000km distance, so at 400km, imagine the strength.)
- In addition, GEO satellites are stationary with respect to the space elevator and will be in the beam (spill-off) of the high-energy laser for long periods of time. As most GEO satellites point at the Earth, the laser is impacting their sensitive nadir side. This could lead to never receiving permission to radiate towards the tether climbers.
- NOTE: Lasers are researched to replace deep space communication as well as satellite communication. The industry and economics of lasers will drive a better system than the stringent demands of the laser clearinghouse sooner than later. Lasers in various strengths are the long-range radio of the future. However, most of these transmissions are low power and not pointed at the GEO arc of satellites.

4.3 Comparison of Power Options

4.3.1 Laser Pros and Cons

The beauty of high-energy lasers is that they can deliver enough energy to climb and conduct operations as the system rises above the surface of the ocean. According to Smith [Smith, 2009], the major advantages of laser-based energy transfer for the space elevator include:

- High power on target
- Each component for the support infrastructure exists; e.g., high power lasers, large telescopes and adaptive optics
- Energy source is from the surface of the planet where it can be operated routinely

The major drawbacks of laser power are:

- Extremely high demands—hundreds of MW needed for each transmitter
- Nuclear power may be needed to deliver sufficient energy
- Power transmitters near the Marine Node will require large laboratory-like facilities and support in the middle of the ocean
- Laser radiation is hazardous, even at low power levels; it can blind people and animals, and at high power levels it can kill through localized spot heating
- Conversion into electricity at the tether climber is inefficient, presenting thermal issues for the power receivers
- Irradiation of the tether climber structure also causes thermal concerns
- Atmospheric absorption and scattering by clouds, fog, rain, etc., causes losses that can be as high as 100%
- Beaming high energy lasers straight up puts significant space assets at risk; the coordination for safe beaming into space is cumbersome and worrisome for non-government entities
- The need for multiple laser sources multiplies these concerns

4.3.2 Solar Power Pros and Cons

The principal rationale for employing the solar power as the source of energy is the elimination of the need to design, develop, deploy and operate a huge infrastructure to create and project energy from the surface of the Earth.

The beauty of the solar array approach is two-fold: (1) there is a tremendous industry with heritage in space technology and, 2) technological research in solar cell improvement from 15% conversion to 28%, to 40%, and even to 70% is ongoing and very positive in the predictions of capability by our need dates (which we estimate as between 2028 and 2030). The advantages of solar powered tether climbers include:

- Power source is continuous with constant output from the sun as long as you are well above LEO

- Above the atmosphere, the physics of the situation is deterministic; the angles to the sun, the amount of eclipse (shadow), the predicted energy received, and the heat absorbed are all known quantities as the Earth rotates with respect to the sun and its seasons
- The solar array technological industry is very large, diverse, global, growing and pushing for efficiency increases
- The space materials industry is similarly robust, active and pushing technology constantly, ensuring that the structural supports for the tether climber's solar array will be much lighter and far stiffer than is possible today, leading to easier deployability after storage and transportation above the atmosphere
- The projected technological capabilities for 2020 will enable sufficient power for tether climbers to operate, especially with the constant power strategy
- Leaving solar arrays at GEO node would enhance capabilities as dual use hardware
- The space elevator operator does not have to build or manage high-power laser transmitters

The major drawbacks for a solar power based energy source include:

- Large solar arrays are required, creating concerns related to size, mass, and structural stiffness
- Solar array efficiency is a critical factor because solar irradiance is low relative to that of laser sources
- Reasonable pointing capability is required that tracks the sun from one horizon to another continuously when there are no eclipses
- Getting to above the atmosphere requires the storing and then deployment of the large and fragile arrays
- Solar array efficiency degrades by about 1.5% per year
- Solar radiation heats the arrays, and this heat will need to be moved and radiated into space to ensure safe operations

4.4 The Hybrid Climber

The ability to adapt to the complexity of space elevator climbing will be important. A suggestion discussed at the 2013 International Space Elevator Conference [Shelef, 2013] deals with the strengths of both laser power and solar power for the energy to lift payloads. The key is flexibility and multiple sources. The initial part of this concept is that once again we need to rise above the atmosphere. The approaches to reach 40 kms altitude were discussed earlier and are expanded upon in Appendix A. Once the tether climber has reached the vacuum of space, the choices are essentially solar power or laser from the ground. These two choices have inherent capabilities that have been discussed previously. They lead to a natural conclusion: lasers early and close to the ground to overcome the large gravity well forces and solar once the tether climber is sufficiently beyond the pull of gravity. The concept is to have the laser receiver cells on one side of the array with the ability to nadir point while on the other side would be solar cells with the ability to point at the sun. This would imply that the large arrays are directional once they switch over to solar. Figure 4-5 shows this concept.

Figure 4-5. Hybrid Climber, Laser down, Solar up (chasedesignstudios.com)

The concept behind this proposal is simple enough. At approximately 40 kms altitude, the photovoltaic unit is unfurled such that the laser side is pointing at the nadir to collect the initial energy needed to prepare for lifting and then climb against the initial 9.8 m/s^2 gravity. The initial speed could be around 40 m/sec with the initial travel about 6,000 kms in about 42 hours. Here are some of the basic calculations associated with this proposal:

- 20 MT * 40 m/sec * g (rounded to 10 m/s^2) = 8MW shaft power
- Assume laser photovoltaic array (LPA) is 75% efficient
- Assume motor is 95% efficient
- Laser power needed is 8 MW / (0.95*0.75) = 11.23MW incident on the laser power receiver
- Due to losses in the atmosphere and beam spillage, an estimate of 20MW of laser power from the ground seems reasonable
- 10kW/m^2 (no more than 10 suns desirable)
- 1,123 m^2 = πr^2. r=~19 m, or 38 m diameter for the laser power receiver array
- If we want to have 2 of them instead (balance/redundancy), each receiver would need to be about 26m in diameter
- Assuming 5kW/kg, 8MW means 1,600kg or 1.6MT for the Laser Power Array

At 6,000 kms, the tether climber would switch over to solar power as the source of energy. The difference here is that the solar array would need to point at the sun to ensure efficient receipt of the energy. The requirement could probably stay at just two axes of control, but further design would have to be accomplished. The numbers that surface for this type of power source are:

Assumptions:
- Gravity is now ~2.5m/s²
- Desired speed to destination is now ~150m/sec
- 20 MT * 150 m/s * 2.5 = 5 MW shaft power needed
- Assume SPA is 40% efficient
- Assume motor is 95% efficient

Solar Array Design estimate:
- 5 MW / (0.95 * 0.4) = 13MW
- Illumination is one sun (1.3kw/m²)
- Array area = 10,000m²
- Works out to 114m in diameter for one array or 80m in diameter for two arrays
- Assuming 5kW/kg of delivered power (see Feasibility Condition for references)
- 5MW / 5kW/kg = 1,000 kg or 1MT

After the design is pulled together for a Hybrid Climber, the size/mass/shape are estimated to be:

- Climber motor – max is 8MW, 95% efficient, 5kW/kg = 1,600 kg (1.6MT)
- LPA = 1.6MT
- SPA = 1MT
- Drive mechanism = 3MT
- Electronics, etc. = 2MT
- Structure = 3MT
- Total = 12.2MT, leaving 7.8MT as payload, about 40%

As the design develops for the tether climber, many options will be considered. The hybrid option could very well become the new baseline. The complexity and added mass for both sources of power may be justified because they enable a very attractive system. One key advantage is it cuts back the power infrastructure at the Marine Node significantly. (Maybe one operational power source on the ground with a backup.)

4.5 Power Source Conclusions

The foregoing analysis leaves us with certain conclusions:

- Transitioning through the atmosphere in a protective box can be enabled with a lightweight power cable.
- Laser energy as the primary source of power for a tether climber ensures the smallest area power receiver among the alternatives.
- Laser energy near the Earth terminus region allows control over energy characteristics (power levels, duration, etc.)

- Solar power is a method with proven techniques utilizing a continuous source of energy (except for short eclipses).
- Solar power receiving cells are rapidly improving in power density (efficiency) and lightweight, foldable structures.
- Solar Power eliminates the need to create energy at the Marine Node.
- The hybrid design offers significant advantages.

References

- Angel, R. and Fugate, R. Science 288:455, 2000.
- Bennett, Harold. 2000. Com-power, private communications, referenced from Edwards 2003.
- Edwards, B. and Westling, E. (2003), "The Space Elevator: A Revolutionary Earth-to-Space Transportation System". BC Edwards, 2003.
- Knapman, John, "Tether Climber at Constant Power," ISEC Conference, Seattle, 2013.
- NASA Tech Briefs, Nov 2012.
- Semon, Ted & Ben Shelef, "Hybrid Climber," a presentation, International Space Elevator Conference, Seattle, Aug 2013.
- Shelef, B. (2008c), "A Solar-Based Space Elevator Architecture," Spaceward Foundation, 2008. Available at: http://www.spaceward.org/elevator-library#SW
- Smith, David (4 January 2009). "Wireless power spells end for cables". The Observer (London).
- Smith, Frederick G., Accetta, Joseph S., and Shumaker, David L., The Infrared and Electro-Optical Systems Handbook: Atmospheric Propagation of Radiation , ANN ARBOR MI , 1993
- USAF (2012), "Energy Horizons", United States Air Force, Energy S&T Vision 2011-2026, AF/ST TR 11-01 31 January 2012, Pgs. 21-24.
- http://spaceelevatorwiki.com/wiki/index.php/SpaceElevatorPower
- http://www.spaceward.org/games

Chapter 5   Conclusions and Recommendations

The International Space Elevator Consortium leveraged many sources to produce this report. Two significant activities led to much of the discussions, then drafts, and finally chapters:

#1 – 2013 International Space Elevator Conference (August 2013 in Seattle). During these three days, focusing on the same topic of space elevator tether climbers, the participants improved upon current concepts and suggested many new ideas. The references show major topics addressed, as a surprising number of new ideas surfaced during the "mini-workshops." The particular workshop on Tether Climbers provided many ideas for this study report. Some of the key ones were:

- Constant Power approach
- Friction discussions (between wheels and tether)
- Hybrid Climber concept
- Solar power option
- Strengths of laser power option
- A new approach for climbing the tether

#2 – Publication of International Academy of Astronautics study report entitled: *Space Elevators: An Assessment of Technological Feasibility and Way Forward.* This four-year study, with 41 authors (many from ISEC), addressed the full spectrum of issues around a space elevator infrastructure. The Academy supported the basic conclusion that space elevators seem feasible. This Academy publication of the study results has provided a significant amount of information on the topic of tether climbers and endorses the movement forward towards an operational space elevator infrastructure.

5.1   General Questions

At the beginning of this report, many questions were posed to stimulate thought and direct analysis. This portion of the chapter will provide some short answers:

- **Is the projected design achievable within the next 15 years?**
  YES
- **Do we know enough to design a basic tether climber?**
  YES, the tether climber is essentially a spacecraft with a special climber apparatus based upon terrestrial designs.
- **What are the special design requirements for the climber?**
  Power sources and the interface between the tether and the climbing wheels.
- **How will the tether climber interact with the tether?**
  The current design has opposing wheels using friction for traction. Externally supplied power will drive the motors that will turn the wheels so that the climber ascends.

- **What are the demands the tether will put on the tether climber?**
  General understanding exists, but until characteristics of the tether are further developed, there is much to be analyzed and assessed.
- **What are the anticipated needs of the developer/owner/operator?**
  6 Metric Ton (MT) climbers, 14 MT payloads, one launch per day, one week trip to GEO, 7.5 years of life for each tether, 10 year MTBF
- **Can we meet these needs?**
  YES
- **Can the tether climber carry its own power source?**
  As the current expected state of the art of materials will not allow the mass to generate and store energy to be raised against gravity, the power must be external.
- **Is solar or laser power preferable?**
  Could be either, or a hybrid of both.

5.2   Conclusions

The following is a set of conclusions reached by the authors of this study.

- The study used the concept of "constant power" as a baseline because of its ability to lower design requirements on the tether climber. This resulted in less mass required in the design and, thus, a more manageable power demand.
- The mass breakout of 6 MT for a climber and 14 MT for customer payloads seems feasible. The estimate is that with a travel time of one week to GEO seven tether climbers can be on a tether simultaneously. Extra tapering of the tether, beyond that needed to support its own weight, can be used to limit the extra demands on the tether from constant power compared with constant speed.
- The development of a full operations concept should be accomplished early to assist in the refinement of requirements.
- The communication architecture should be integrated into the space elevator infrastructure and nodal layout. This will enable the tether climber to be in constant contact with operators and customers.
- Solar power, as the sole source, appears achievable and would eliminate the need to create and operate large energy sources on the surface.
- The use of laser power as the sole source also seems achievable.
- Detailed designs of the climber, tether attachment apparatus, and drive mechanisms will be accomplished later when the characteristics of the tether are better defined. There will be a great deal of leverage from terrestrial motor design to leverage climber development.

5.3   Recommendations

Recommendation 1. Further ISEC study must be accomplished while keeping up with the following technologies:

- CNT tether design
- Solar array development
- Laser projection capability
- Lightweight spacecraft materials
- High stage one
- Linear motor design
- Lightweight space-capable batteries

Recommendation 2. ISEC should continue its support of tether climber and tether strength competitions around the world. In addition, support to design exercises for school students of all ages (for example, the Seattle Museum of Flight's Lego competitions) must be continued to ensure the next generation stays interested in space elevators.

Recommendation 3. ISEC should continue and strengthen its international relationships, including those with the Japanese Space Elevator Association (JSEA), the Eurospaceward Foundation, and the International Academy of Astronautics.

Appendix A: Movement to 40 Kms

(Note: Much of this appendix comes from "Space Elevators: An Assessment of the Technological Feasibility and Way Forward," [Swan, 2013])

All space elevator tether climbers will leverage the ability to separate the engineering requirements (design criteria) into two significant phases. The first phase will be designed to safely transit the first 40 km from the surface of the ocean. This will require protection from massive winds, lightning and ice. The protection from these atmospheric environmental threats will require mass in the form of protective devices. The simple, preliminary design is a "box" that surrounds the tether climber and all its equipment, to include solar arrays or laser receiver arrays. This protective box will have significant structure and mass.

The second phase would enable travel from 40 km to the Apex Anchor. Indeed, there will be environmental threats that must be designed for, but will probably not need the massive wind protection or lightning isolation mass. Thus, 99% of the travel of tether climbers can be accomplished without "raising against gravity" mass only needed for 1% of the trip. This leads to the conclusion that the tether climber is designed for conditions above 40 km with a protective box for transit to 40 km altitude. At the present time, there are three approaches for the transit to altitude. They are described below:

A.1. Marine Stage One – Box Protection with power cable

The approach here is that the movement of the tether climber and its power source – solar arrays or laser receivers. This would all be placed inside a protective box and raised to an appropriate altitude. The current concept for tether climbers with the solar power option is that, above 40km the system will only climb while in sunlight and hibernate during eclipse. The laser power option would not have this restriction. Some discussion points are as follows:

Starting Point: Ocean surface

Above Atmosphere: Assumed to be 40km (TBD after further study)

Box Concept: The box would be made from a very light material that would totally encompass the tether climber and its solar cells so that all the effects of the atmosphere would be kept from the climber system. The idea is that the major wind forces, the lightning, etc., would be denied access to the vulnerable climber and its solar cells. The size, shape and material will be determined after much study into the future needs of tether climbers. Assumption: 10 Metric tons.

Power: During the ascent from ocean surface to the appropriate height [40km], the climber and its customer payload would have power supplied by a lightweight carbon nanotube power cable. An alternative could be localized laser power; however, this is not the preferred approach.

Speed: The requirement is that the climber be free of the protective box prior to the dawn of the first day, when power could be applied to the solar cells of the climber.

Post Release: Once the tether climber is released from the protective box, the box and all its support equipment will be returned to the surface, for re-use. The total mass of this protective box must be calculated as part of the total load of the space elevator and could impact the number of climbers on the tether at any one time or, more likely, lead to a requirement for a stronger tether.

Propulsion: Currently, the thinking is that the same motor and engine drive could be used to propel the protective box and its payload [tether climber and solar cells] to the 40km altitude. However, a different drive will be needed if atmospheric conditions lead to the requirement for a different tether shape in the atmosphere, perhaps more like a rope rather than a broad ribbon. Alternatives could include rockets, balloons to a certain altitude and then another means of propulsion. Also, counter weights could be leveraged from a higher location on the tether of the space elevator.

Return: Currently, the thinking is that the protective box is reusable, but this could be argued if the benefits out-weigh the negatives with "throw-away" boxes.

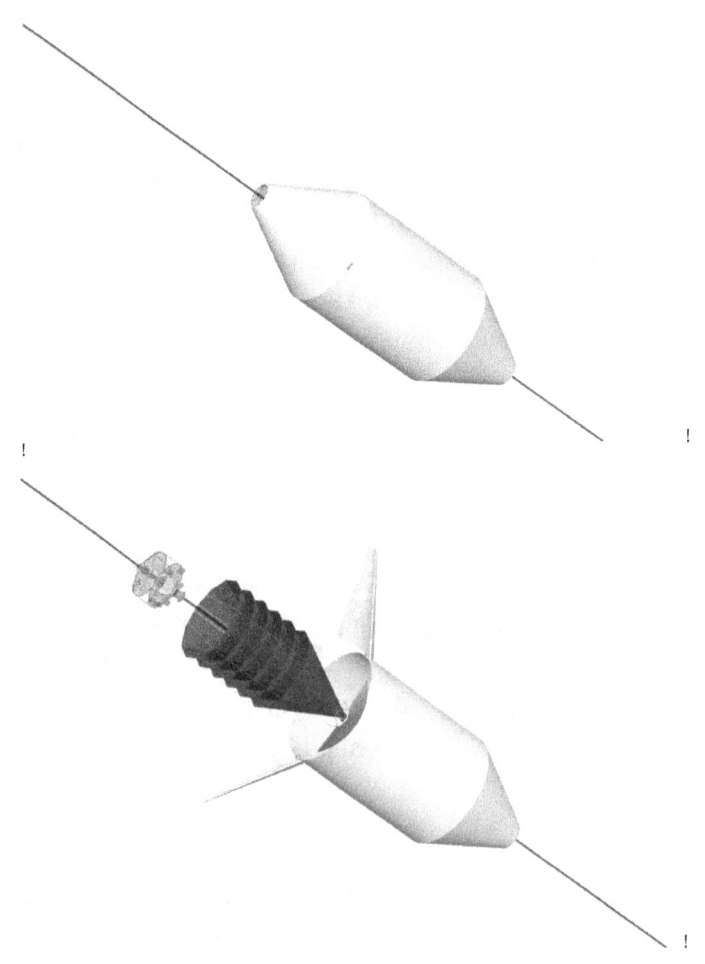

Figure A-1. Box Protection Satellite (chasedesignstudios.com)

A.2. Marine Stage One – Spring Forward

This approach for the initiation of tether climbers is one that takes advantage of the characteristics of the tether and leverages the routineness of transportation infrastructures. The process is conceptually simple:

- Reel in the tether
- Attach tether climber (with protective box covering)
- Allow tether to pull the tether climber beyond 40km in altitude

The concept comes from the basics of a very long springy material that has stretch characteristics of approximately 0.1% or 1m in each kilometer. This is an estimate and will have to be assessed after the material is developed for the space elevator. When the space elevator is 100,000km long, the natural stretch of the material is approximately 100,000m or 100km. Leveraging this material's stretching capability, the space elevator can easily handle the first 40km of altitude movement up and down. After the space elevator is established in its operational position, the machinery can be established to rapidly reel in the tether and then reel out the tether with climber. The objectives of this approach, called "Spring Forward" are:

- Avoid the atmospheric effects on the tether climber by sending it up inside a protective box.
- Keep operations simple
- Have power during activities
- Launch tether climbers once per day
- Enable the solar arrays to be deployed at altitude

Approach for Spring Forward: The primary purpose of this approach to launching space elevator climbers is to enable the first 40km to be traversed with a shield for the environment. The shield would then be returned to the surface as the tether climber is kick-started for GEO and beyond with solar arrays deployed. (This approach uses solar power as an example.) This enables the designers to negate the effects of the atmosphere and allow the solar arrays to be fragile, large and efficient. This would enable the space elevator climbers to operate ONLY on solar power after the 40km altitude release from the shield. This ability to design the solar arrays for "no winds" enables a simple design for the tether climbers enabling a raise from 40km to 100,000km. The approach to enable this concept is relatively simple:

Step 1: Reel In – A winch reels in the tether to the base station at sea during the night. The previous climber was released for climb at daybreak at altitude [6 am roughly], so we need to delay reeling in as long as possible to avoid pulling the previous climber too low. (The lower it is, the heavier it gets.)

Step 2: Attachment – After the tether has been reeled in, the next space elevator tether climber, boxed for ascent, will be aligned at the base station with the tether. In addition, a lightweight electrical cable will be attached to the climber to provide power for the first 40km until sunrise.

Step 3: Release and Climb – This step occurs when the tether climber has been prepared for lift and climb with its power cable attached, its solar arrays attached and the operations center approval. As the ribbon is reacting to the stretch, the tether climber will rise up to the altitude that is above the major effects of the atmosphere while being protected by the shield.

Step 4: Preparation for Climb – Once the tether has pulled the tether climber up above the atmospheric effects, the climber must be prepared for its long ascent to GEO and beyond. The first step is to pull the climber out of the shield while deploying the solar arrays. The current concept is that it would be stacked and then pulled vertically out of the cylinder until the solar arrays are hanging down from the climber pointed towards the direction of the rising sun. Meanwhile, the power cable is providing the necessary energy for heat and preparation for climb. There will be constant communication with mission control to ensure everything is ready for the seven-day climb.

Step 5: Checkout – This is a critical step in that the climber will be released for ascent when approved from operations center. As the sun rises, the solar arrays will provide the energy to the batteries and the climber motor. When everything is ready, the tether climber will be released for climb.

Step 6: Climb – This is the actual mission of the space elevator infrastructure. Initiation of ascent , outside of the atmosphere, enables the design team to use only solar arrays for power sources.

Shield Design: The key here is that the package [tether climber and solar arrays] must be protected during the spring up to 40 km from the surface of the ocean on the marine platform. The simple concept is to have a cylindrical shaped shield be wide enough to handle the package in its compact form at zero altitude and long enough to include both the tether climber and the stacked solar arrays. The initial concept is that the shield would be a permanent cylinder attached to the space elevator tether at its 40 km altitude location. It is then pulled down to the marine platform—stretching the tether—where it would be opened and loaded. The shield would have to include protection from all the aspects of the atmosphere to include massive winds, lightning, ice, upper atmospheric static charges, and other effects as they are understood. The climber, once it reaches the appropriate altitude, would then climb vertically out of the shield and extend its solar arrays, and depart on its mission.

A.3. High Stage One

A second option in location of the space elevator Earth terminus takes the complexity and weight of traveling through the atmosphere off the space elevator tether and places it on an Earth-based structure 40 km high. The stresses induced by the lower and upper atmospheres are dealt with by infrastructure based firmly on the Earth's surface. The space elevator is able to deal with the effects of Earth's turbulent atmosphere without adding to the weight that has to be supported from geosynchronous orbit. High Stage One achieves this by keeping the tether in and above the mesosphere

The concept is simple as shown in figures A-2 and A-3: place the working end of the space elevator on a firm platform at altitude. This facility would be capable of supporting 3,000 tons at 40km altitude with NO forces on the space-elevator tether. As such, stresses on the tether become only space-oriented forces, not atmospheric forces. The Lofstrom Loop [Lofstrom 1985] was proposed as an alternative high-altitude launch location. It ensures stability of the platform at altitude and provides routine access from the ocean surface to 40 km altitude using electric trams. This transfer of hazards and forces from the lower portion of the space elevator infrastructure to the surface based Lofstrom Loop simplifies the problem and reduces the mass requirement of the space elevator tether by a factor of 10. Once the platform has been established at 40 km altitude and the logistics "train" has geared up, the space elevator infrastructure becomes safer and simpler. The concept of High Stage One using the Lofstrom Loop is shown in the next two figures.

Figure A-2. Transfer Platform [Knapman, 2013]

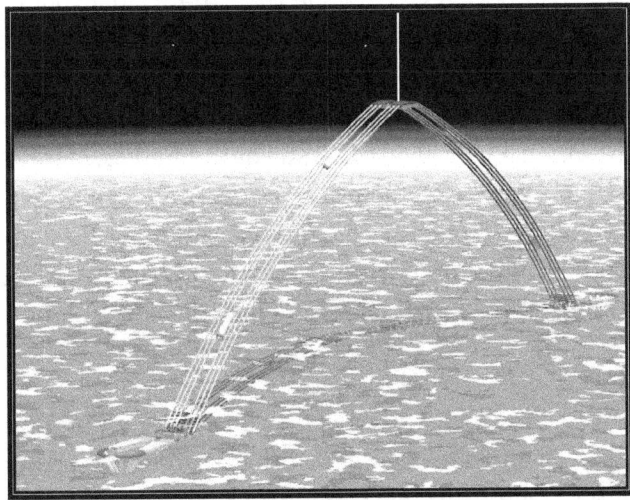

Figure A-3. High Stage One with surface stations and transfer platform [Knapman, 2013]

Appendix B: Tether Baseline

The systems engineering approach for complex systems enables piecewise approaches to huge projects. In this case, the study is focusing on the tether climber and needs to have a baseline space elevator tether as something to design against. Space elevator engineers have determined that the material is not available, even in a laboratory, that could meet all the requirements. As this study will not address this issue, the study lays out what the tether climber needs in order to become an effective space elevator component. From past activities and studies, the expected characteristics for a tether would be:

Length – 100,000 km to Apex Anchor

Width – 1 meter, curved

Approach – woven strands of CNTs that build from molecular strands, to small strings, to small ropes, and then to longer and longer elements of a long woven CNT tether.

Load Carry – 31+ Metric Ton single load at Earth terminal. The load depends on which method is selected to get climbers to 40km altitude.

Strength of Material – 25-30 MYuri specific strength would suffice; this study will baseline 27 MYuri. A 38 MYuri material provides 27 MYuri with 1.4 safety factor.

Taper – As the material will not support the total load as a simple tether, there must be tapering along the tether growing from the surface of the ocean (minimum load) to the maximum load at GEO. For a 27 MYuri capable tether, the taper is 6 if tether climbers travel at constant speed; it is 8 if the climbers travel at constant power.

Taper Approach – Two approaches are available. The first is just to increase the width as one rises, up to a width of 6 or 8 meters at GEO, while the other is just to increase the thickness of the material in the 1-meter wide tether. This report chooses the latter as that gives a constant cross-section for the climber designers. The current estimate is that the tether material immediately above the atmosphere would be 1-meter wide by 10.5 microns thick (8.3 microns with climbers at constant power) to provide the strength needed. As a result, the GEO size would either be 6meters wide with 10.5 microns thickness (8.3 meters by 8.3 microns with constant power) or 1-meter wide by 63 (or 69) microns thick. If one were to set up the one-meter wide tether with ten strands, they would be 0.2 cm wide and 525 (or 415) microns thick, while at GEO there would be 60 (or 83) of the same strands in the same cross section of one meter.

Environmental Effects – This arena of study will be quite detailed in other work. One key is that the threats vary depending on altitude region, so the design of the tether should vary as the environment changes due to winds, atomic oxygen, electrodynamic and magnetic interactions, as well as space debris and meteors. These will probably not affect the design of the tether climber and so will not be addressed inside this report.

Lifetime – ten years (planned replacement on 7.5 years)

Loading on Tether – Seven climbers – one week to GEO – 20 Metric Tons mass each

Availability of Tether – 2030 in 100,000 km length

Table B-1 provides a summary of these characteristics.

Tether rated strength	38 MYuri (49.4 GPa)
Operational load	27 MYuri (35.2 GPa)
Safety factor	1.4× (40%)
Density of tether material	1.3 gm/cm^3
Taper ratio	6 or 8
Cross section at Earth terminus	10.5 or 8.3 mm^2 (1m wide × 10.5 or 8.3 µm)
Cross section at GEO	63 or 69 mm^2 (1m wide × 62.8 or 68.6 µm)
Tether climber mass	20 MT (6 MT structure, 14 MT payload)
Number of Climbers on tether simultaneously	7
Equivalent climber weight (after accounting for effects of reduced gravity at altitude)	29 MT
Tether Mass	6,300 or 6,900 MT
Apex Anchor	1,900 or 2,100 MT (30% of tether mass)
Tether Length	100,000 km

Table B-1. Tether characteristics

Appendix C: An Alternate Climbing Approach

During the mini-workshop on Tether Climbers at the August 2013 International Space Elevator Conference in Seattle, a unique idea surfaced during the brainstorming, proposed by Dalong "Sam" An and Peter N. Glaskowsky.

A major concern recognized by space elevator designers is a lack of knowledge of the surface characteristics of a tether made from carbon nanotubes. Uncertainty regarding coefficients of friction, wear rates, and other key factors makes it difficult to determine the final design characteristics of the tether climber drive system.

The current baseline for the interface between the tether and the drive system assumes that pairs of metal-surfaced wheels will drive directly against a one-meter wide CNT woven tether ribbon. This direct-drive method has been a baseline approach for over 13 years and has not been challenged. It is clear, however, that if a CNT ribbon exhibits low surface friction, direct drive may require extremely high levels of normal force between the drive wheels and the ribbon to create enough friction to support the 20 MT weight of the tether climber and its payload. Managing these forces may require extremely strong structures, wheel bearings, and other components, adding significant weight to the tether climber, and the forces may cause significant wear on the tether itself.

In looking for alternatives to the direct-drive method that might reduce structural complexity and tether wear, a new system for tether climbing was proposed. The system consists of three objects in series on the tether—the tether climber itself, and two powered *shuttles* above it. A loop of commercial off-the-shelf rope goes through all three objects, hanging parallel to the tether.

Each shuttle has a pair of clamps, allowing it to grip the tether, the rope, or both, as well as a small drive system that can pull the shuttle up the tether. While ascending, a shuttle carries only the load of its own weight plus (for the upper shuttle only) the weight of the rope.

Inside the tether climber, the loop of rope passes through a motor-driven capstan that allows the climber to ascend the rope at some velocity V. This is the only drive system that needs to be able to lift the full weight of the tether climber and its payload.

The shuttles operate in a four-phase cycle:

- Phase 1. The lower shuttle clamps the tether and the rope, supporting the weight of the tether climber. The upper shuttle, which is not clamped to the tether or rope, engages its tether drive system and climbs the tether at a velocity of *2V*, lifting only the weight of itself and the rope loop.
- Phase 2. When the upper shuttle reaches the limit of the rope loop, it clamps the tether and rope and disengages its tether drive system. The lower shuttle unclamps the tether

and rope, thus transferring the weight of the ascending tether climber to the upper shuttle.

Phase 3. The lower shuttle engages its tether drive system and starts climbing the tether at velocity *2V*, lifting only its own weight.

Phase 4. When the lower shuttle reaches the upper shuttle, it clamps the tether and rope and disengages its tether drive system. The upper shuttle unclamps the tether and rope, thus returning the system to the conditions of Phase 1.

Throughout all four of these phases, the tether climber is continuously climbing the rope using its capstan drive system. The tether climber has its own tether clamps to be used when the climber is stopped (when deploying a payload, for example), but these clamps will be disengaged while the climber is ascending. Note that in practice, the shuttles will have to ascend at a speed slightly higher than *2V* to allow for the time needed to verify safe transitions between the four phases of operation.

Commercial ropes are readily available that meet the strength requirements for this system. For example, a 2" Amsteel-Blue Dyneema rope from Samson Rope has a minimum strength of 156 metric tons[1]. Ropes of this type are used to moor and tow large ships, in mining equipment, and for other demanding applications, often in conjunction with capstan drives. A 1-km length of this rope has a mass of 1.29 MT, a relatively small figure that may compare well to the structural weight savings enabled by this drive system. Of course, other materials (such as stranded or solid metallic or carbon-composite belts[2]) could also be used.

The side of the rope loop nearest the tether (the "tight" side) remains stationary relative to the tether because at least one shuttle is always clamped to both the tether and the rope. The other side of the rope loop (the "loose" side) moves intermittently upward. Because the loose side of the rope rises only when the upper shuttle is climbing, it must move at over four times the average speed of the climber, and so will need to be managed carefully to keep it from whipping around. When the upper shuttle is clamped and the lower shuttle is moving, the loose side of the rope just hangs in place.

It will likely be desirable to provide redundancy by equipping the system with two rope loops and two sets of clamps and drive systems on the shuttles and tether climber, for example on opposite sides of the tether ribbon.

---

[1] http://www.samsonrope.com/Pages/Product.aspx?ProductID=872
[2] http://www.kone.com/ultrarope

Appendix D: The International Space Elevator Consortium

Who We Are

The International Space Elevator Consortium (ISEC) is composed of individuals and organizations from around the world who share a vision of humanity in space.

Our Vision

A world with inexpensive, safe, routine, and efficient access to space for the benefit of all mankind.

Our Mission

The ISEC promotes the development, construction and operation of a space elevator infrastructure as a revolutionary and efficient way to space for all humanity.

What We Do

Provide technical leadership promoting development, construction, and operation of space elevator infrastructures.

Become the "go to" organization for all things space elevator.

Energize and stimulate the public and the space community to support a space elevator for low-cost access to space.

Stimulate science, technology, engineering, and mathematics (STEM) educational activities while supporting educational gatherings, meetings, workshops, classes, and other similar events to carry out this mission.

A Brief History of ISEC

The idea for an organization like ISEC had been discussed for years, but it wasn't until the Space Elevator Conference in Redmond, Washington, in July of 2008, that things became serious. Interest and enthusiasm for the space elevator had reached an all-time peak, and with Space Elevator conferences upcoming in both Europe and Japan, it was felt that this was the time to formalize an international organization. An initial set of directors and officers were elected, and they immediately began the difficult task of unifying the disparate efforts of space elevator supporters worldwide.

ISEC's first Strategic Plan was adopted in January of 2010 and it is now the driving force behind ISEC's efforts. This Strategic Plan calls for adopting a yearly theme to focus ISEC activities. (For 2010, the theme was "Space Elevator Survivability - Space Debris Mitigation.") In 2010, ISEC also announced the first annual Artsutanov and Pearson prizes to be awarded for "exceptional papers that advance our understanding of the Space Elevator."

Because of our common goals and hopes for the future of mankind off-planet, ISEC became an Affiliate of the National Space Society in August of 2013.

Our Approach

ISEC's activities are pushing the concept of space elevators forward. These cross all the disciplines and encourage people from around the world to participate. The following activities are being accomplished in parallel:

CLIMB – This annual peer-reviewed journal invites and evaluates papers and presents them in an annual publication with the purpose of explaining technical advances to the public. The first issue of CLIMB was dedicated to Mr. Yuri Artsutanov (a co-inventor of the space elevator concept), and the second issue was dedicated to Mr. Jerome Pearson (another co-inventor). CLIMB is scheduled for publication in July each year.

Yearly conference – International space elevator conferences were initiated by Dr. Brad Edwards in the Seattle area in 2002. Follow-on conferences were in Santa Fe (2003), Washington DC (2004), Albuquerque (2005/6 –smaller sessions), and Seattle (2008 to the present). At each of these conferences, there were many discussions across the whole arena of space elevators with remarkable concepts and presentations. Recent conferences have been sponsored by Microsoft, the Seattle Museum of Flight, the Space Elevator Blog, the Leeward Space Foundation, and ISEC.

Yearlong technical studies – ISEC sponsors research into a focused topic each year to ensure progress in a discipline within the space elevator project. The first such study was conducted in 2010 to evaluate the threat of space debris. The second study, and resulting report, focused on space elevator operations. The current study is focusing on tether climber designs. The 2014 topic has been chosen: Space Elevator Architectures and Roadmaps. The products from these studies are reports that are published to document progress in the development of space elevators.

International cooperation – The ISEC supports many activities around the globe to ensure that space elevators keep progressing towards a developmental program. International activities include coordinating with the two other major societies focusing on space elevators: the Japanese Space Elevator Association and EuroSpaceward. In addition, ISEC supports symposia and presentations at the International Academy of Astronautics and the International Astronautical Federation Congress each year.

Competitions – ISEC has a history of actively supporting competitions that push technologies in the area of space elevators. The initial activities were centered on NASA's Centennial Challenges called "Elevator: 2010." Inside this were two specific challenges: Tether Challenge and Beam Power Challenge. The highlight came when Laser Motive won $900,000 in 2009, as they reached one kilometer in altitude racing other teams up a tether suspended from a helicopter. There were also multiple competitions where different strengths of materials were tested going for a NASA prize – with no winners. In addition, ISEC supports educational efforts of various organizations, such as the LEGO space elevator climb competition, at our Seattle conference. Competitions have also been conducted in both Japan and Europe.

Publications – ISEC publishes a monthly e-Newsletter, its yearly study reports and an annual technical journal [CLIMB] to help spread information about space elevators.

Reference material – ISEC is building a Space Elevator Library, including a reference database of Space Elevator related papers and publications.

Outreach – People need to be made aware of the idea of a space elevator. Our outreach activity is responsible for providing the blueprint to reach societal, governmental, educational, and media institutions and expose them to the benefits of space elevators. ISEC members are readily available to speak at conferences and other public events in support of the space elevator. In addition to our monthly e-Newsletter, we are also on Facebook, Linked In, and Twitter.

Legal – The space elevator is going to break new legal ground. Existing space treaties may need to be amended. New treaties may be needed. International cooperation must be sought. Insurability will be a requirement. Legal activities encompass the legal foundation of the space elevator - international maritime, air, and space law. Also, there will be interest within intellectual property, liability, and commerce law. Starting work on the legal foundation well in advance will result in a more rational product - legal opinions that pre-date a space elevator.

History Committee – The ISEC supports a small group of volunteers to document history of the space elevator. The committee's purpose is to provide insight into the progress being achieved currently and over the last century.

Research Committee – The ISEC is gathering the insight of researchers from around the world with respect to the future of space elevators. As scientific papers, reports and books are published, the research committee is pulling together the relative progress to assist academia and industry to progress towards an operational space elevator infrastructure.

ISEC is a traditional not-for-profit 501 (c) (3) organization, with a board of directors and four officers: President, Vice President, Treasurer, and Secretary. In addition, ISEC is closely associated with the conference preparation team and other volunteer members.

Address: ISEC, 709-A N. Shoreline Blvd, Mountain View, CA 94043
(630) 240-4797 / inbox@isec.org / www.isec.org

www.ingramcontent.com/pod-product-compliance
Lightning Source LLC
Chambersburg PA
CBHW080945170526
45158CB00008B/2381